First published in the United States by
The Lighthouse Academy Press
Printed by Kindle Direct Publishing, USA
Additional copies for sale at Amazon Books

ISBN: 979-8-9953537-1-3

This work was created through a human–AI collaboration. The author directed, revised, and finalized all content. See *Author's Note.*

What is Called Artificial Intelligence?

William J. Striker

Contents

Author's Note

This work began not as a response to the current discussion of artificial intelligence, but as a response to its confusion.

The term itself has become common currency — used in technical circles, policy debates, marketing language, and ordinary conversation — yet rarely defined with precision. Systems of very different kinds are described under a single name. Capabilities are assumed from appearances. Limitations are ignored. The result is not merely disagreement, but misplacement: arguments proceeding without a clear understanding of the subject itself.

This book was written to restore that footing.

It does so by returning to first principles — beginning with definition, then tracing development, then examining use, and finally addressing fear. The aim is not to speculate beyond what is known, but to bring what is known into clearer form. Artificial intelligence is neither trivial nor mysterious. It is the product of identifiable forces, developed over time, and used in ways that can be observed and evaluated.

A word is necessary regarding authorship.

This work was developed with the assistance of modern AI systems. These systems were used as instruments of exploration — to test structure, examine phrasing, and assist in organizing complex material. They contributed to drafting, but not to the judgment that shaped the final work.

The selection of concepts, the framing of arguments, the acceptance or rejection of ideas, and the responsibility for every conclusion remain human.

Artificial intelligence, as described in this book, does not think. It does not judge from within a point of view. It does not hold meaning. It produces outputs through learned patterns applied to input. That capability can be powerful, but it is not authorship.

Authorship requires judgment.

For that reason, this work should be read as a human work, assisted by artificial means. The tools used in its construction reflect the subject of the book itself: the interaction between human judgment and engineered systems. But the voice, the direction, and the responsibility remain singular.

Preface

An explanation of the title is in order:

What is Called Artificial Intelligence?

The term *artificial intelligence* now appears everywhere — in headlines, boardrooms, policy papers, technical marketing, and ordinary conversation. Yet frequency of use has not produced precision. The phrase is applied to systems of very different kinds, and discussion often proceeds as though computation, intelligence, thinking, judgment, and autonomy were interchangeable. They are not.

This book begins, therefore, where serious discussion must begin: with definition. As Voltaire observed, if we are to converse meaningfully, we must first define our terms. Much of the present confusion surrounding artificial intelligence arises from failing to do so. Systems are praised or feared before they are clearly described. Capabilities are inferred from appearances. Limits are ignored. Names begin to do the work that analysis should do.

But definition alone is not enough. To understand what artificial intelligence is, one must also understand how it became possible, how it is used, and why it provokes such strong reactions. For that reason, the book proceeds in three parts.

Part I addresses what artificial intelligence is, and how it came to be. Artificial intelligence did not arrive — it accumulated. What is now called AI emerged through the long interaction of computing power and programming power, until systems capable of learned pattern use became practical at scale.

Part II addresses how artificial intelligence enters society. AI is not one tool but a class of tools, and its effects are best understood through two primary functions: convenience and productivity. These two modes shape behavior, work, expectations, and the place of human judgment within institutions.

Part III addresses why artificial intelligence provokes fear. Some concerns are warranted. Others are projections. The central task is to separate them, distinguishing between real risks such as human misuse, over-delegation, and institutional distortion, from speculative fears that assign thought, intention, or inwardness where none has been shown.

The title itself is a deliberate nod to Martin Heidegger's *What Is Called Thinking?* The question here is not merely what artificial intelligence *is*, but what is meant when we call something *artificial intelligence*. Names carry assumptions, and assumptions, when left unexamined, begin to govern conclusions.

This book does not attempt to settle every question that may one day arise around artificial intelligence. It aims, more modestly and more usefully, to establish a firm footing from which such questions can be asked clearly and judged soberly.

What is Called Artificial Intelligence?

Introduction

The Problem of Misplaced Attribution

Artificial intelligence now occupies a central place in modern life, yet the term remains unstable. It is used to describe systems of very different kinds and capabilities, often with little distinction between computation, learned adaptation, thinking, judgment, and autonomy. The result is a peculiar confusion: AI is at once overstated, understated, and misunderstood.

A simpler footing is available. Current artificial intelligence is best understood as a class of engineered adaptive systems whose useful behavior depends on learned structure rather than solely on explicit rule chains. Such systems can be intelligent in a functional sense without thereby thinking, understanding, or judging from within a point of view.

From that basis, three connected claims follow.

Modern AI did not appear as a single invention or sudden rupture. It emerged through the long interaction of computing power and programming power, together with the methods that made those capacities usable.

AI enters society chiefly through convenience and productivity, and those two modes produce distinct effects on behavior, work, and institutional life.

The greatest present dangers do not arise from machine inwardness, but from human misuse, over-delegation, and the displacement of judgment by speed, scale, and institutional abstraction.

The task, then, is neither to mythologize artificial intelligence nor to dismiss it, but to describe it clearly enough that its capabilities, limits, and risks can be judged in their proper place. The question is not what fantasies can be projected onto AI, but what kind of system it actually is, how it became possible, and what follows from using it as we now do.

Part I:
What is Called
Artificial Intelligence?

Chapter 1: What is Called Artificial Intelligence? Modern Meaning

The phrase *artificial intelligence* did not arise recently. It was introduced in 1956 at the Dartmouth Conference, where a small group of researchers gathered to explore the possibility that aspects of human intelligence might be simulated by machines.[1]

The term itself is revealing. It was not discovered. It was proposed.

From the beginning, artificial intelligence was a **name applied to an ambition** — the attempt to construct systems capable of performing tasks which, if performed by humans, would be considered intelligent. That origin is important. The field did not begin with a clear object called "artificial intelligence." It began with a hypothesis and a label.

Since that time, the meaning of the term has shifted.

The phrase is now used so broadly that it often conceals more than it clarifies. It is applied to search engines, fraud detection, recommendation systems, autonomous vehicles, chatbots, image generators, and even to ordinary software that

---

[1] The Dartmouth Summer Research Project on Artificial Intelligence (1956), organized primarily by John McCarthy, Marvin Minsky, Nathaniel Rochester, and Claude Shannon, is widely regarded as the founding event of the field. The proposal assumed that "every aspect of learning or any other feature of intelligence can in principle be so precisely described that a machine can be made to simulate it". The significance of the conference lies not in a technical breakthrough, but in the naming and framing of the problem itself — artificial intelligence as an object of systematic study.

has merely been touched in some way by machine learning. These are not all the same thing.

For present purposes, the modern usage is the one that matters.

Artificial intelligence (AI) refers most usefully to a class of engineered systems whose behavior depends substantially on **learned structure** — patterns derived from data, statistical relationships, and optimization — rather than solely on explicit chains of rules written in advance.

These systems ingest data, detect structure, and apply that structure to new inputs in order to produce outputs. Those outputs may take the form of prediction, classification, ranking, recommendation, decision support, or synthetic content such as text, images, audio, or video. Many of the systems now dominant in public discussion are also *foundation models* — broad models trained once and adapted across many tasks.

This is the practical territory of modern AI.

What is Meant by "Artificial"

The adjective *artificial* is not decorative. It points first to origin.

These systems are artificial because they are constructed. Their architecture, training data, optimization objectives, and deployment context are all products of human design. Whatever capability they exhibit appears within a system built for that purpose.

The word also preserves an important boundary.

Artificial intelligence is not natural intelligence transplanted into silicon. It is an engineered analogue that reproduces certain outward effects associated with intelligence while leaving open the deeper questions of understanding, consciousness, intention, and awareness.

A system may produce an output that appears intelligent without settling whether it thinks in any human sense.

From Explicit Instructions to Learned Behavior

Traditional software executes explicit instructions.

The programmer defines the logic step by step, and the machine carries it out. If inputs and system state remain the same, the result is expected to be the same. The behavior of the system is specified in advance.

Modern AI systems differ in a crucial way.

Their useful behavior is often not written out line by line. Instead, the system is trained. It is exposed to examples, adjusted through optimization, and shaped by data until it produces useful outputs under new conditions.

The machine still computes. Nothing magical has been added.

But what it computes is now governed by learned structure rather than wholly explicit instruction. The behavior of the system reflects patterns discovered during training, not only rules written beforehand.

This is the essential shift.

When Do We Call It Artificial Intelligence?

Not every computational system is properly described as artificial intelligence.

A payroll system, a database engine, or a structural analysis program may be highly sophisticated, but its behavior is determined in advance through explicit procedure. These are not AI in the present sense.

The term becomes appropriate when useful behavior is no longer fully prescribed.

Instead, the system acquires structure from data and applies that structure across variation. Its responses are not confined to predefined cases, but extend through learned patterns.

This capacity for learned adaptation marks a clear departure from traditional software.

Artificial intelligence is therefore not a single technology, but a class of systems in which learning and adaptability play a central role in producing useful outcomes.

The Present Public Face of AI — Language Models

Much of today's public discussion of AI is, in practice, a discussion of large language models.

These systems are trained on vast bodies of text so that they learn the statistical structure of language. When used, they generate output step by step by estimating likely continuations given the context already provided.

This gives rise to a common misunderstanding.

Because the output is fluent, it is often taken as evidence of understanding. But fluency is not proof of comprehension. Persuasive language is not proof of thought.

A system may be highly effective at producing coherent language without possessing awareness of what it says.[2]

---

[2] An early illustration of this distinction appears in Appendix E, where coherent technical language is generated through structured pattern selection without underlying understanding.

A Working Definition

For the purposes of this book, artificial intelligence will mean:

> Engineered systems that produce **adaptive-seeming outputs** by applying **learned patterns, statistical relationships, or model-based structure** rather than relying solely on fixed, explicit rule chains.

This definition is deliberately functional.

It allows present-day AI to be described clearly while withholding larger claims than the evidence supports. It does not assume consciousness, self-awareness, intention, or understanding in the human sense.

It simply describes what these systems do.

Sources and Notes

1. National Institute of Standards and Technology. (2023). *Artificial Intelligence Risk Management Framework (AI RMF 1.0).* Useful for a modern, functional description of AI systems in terms of outputs such as predictions, recommendations, and decisions.

2. Russell, S., & Norvig, P. (2021). *Artificial Intelligence: A Modern Approach* (4th ed.). Pearson.
   Standard technical overview of the field, especially useful for distinguishing AI as a broad discipline from the narrower public meaning now dominant.

3. Mitchell, M. (2019). *Artificial Intelligence: A Guide for Thinking Humans.* Farrar, Straus and Giroux.
   A clear and careful discussion of what present-day AI systems do well, where they fail, and why fluent output should not be confused with human understanding.

4. Goodfellow, I., Bengio, Y., & Courville, A. (2016). *Deep Learning.* MIT Press.
   Foundational treatment of learned parameters, training, optimization, and model behavior.

5. Vaswani, A., et al. (2017). "Attention Is All You Need."
   Seminal paper on the transformer architecture underlying most modern large language models.

6. Brown, T. B., et al. (2020). "Language Models are Few-Shot Learners."
   Landmark demonstration of scale effects in large language models.

7. Bommasani, R., et al. (2021). "On the Opportunities and Risks of Foundation Models."
   Helpful for describing broad, reusable models trained on large and diverse data sets.

8. *Oxford English Dictionary.* "artificial."
   Useful for preserving the plain meaning of the adjective and avoiding unnecessary mystification.

9. Ji, Z., et al. (2023). "Survey of Hallucination in Natural Language Generation." *ACM Computing Surveys.*
   Useful for understanding the distinctive failure mode of fluent but false generated output.

Chapter 2:
What Computers Do vs What AI Does

The title of this chapter invites a comparison, but a caution is needed at the outset. Artificial intelligence does not operate outside the computer, nor does it transcend computation. AI runs on computers, and every AI system remains, at bottom, a computational system.

The difference, therefore, is not between one kind of machine and another, as though the ordinary computer calculates while AI does something wholly different. The difference lies in how the machine is made useful — how inputs are transformed into outputs, where the effective behavior of the system comes from, and what kind of result the system is expected to produce.

This matters because much of the present confusion begins precisely here. If it is not made clearly, machines are attributed powers they do not possess, while the capabilities they do possess are misunderstood.

Computation as the Execution of Explicit Procedure

A computer, in the classical sense, executes instructions.

It stores data, retrieves it, compares it, sorts it, transforms it, and carries out arithmetic and logical operations according to rules defined in advance. The essential characteristic is not speed alone, though speed matters. It is fidelity to procedure.

In such systems, the programmer specifies the route from input to output. If one writes a payroll system, a structural analysis routine, a tax calculator, or a database query engine, the logic is laid down beforehand. The machine then executes that logic with speed and consistency. If the same inputs and system state are presented again, the same result is expected.

This form of computation remains the backbone of modern civilization. Financial systems, air traffic management, encryption, inventory control, spreadsheets, search indexes, operating systems, and large portions of engineering analysis all depend on explicit procedure. These are not primitive forms of computing. They are the expressions of reliability.

When the rules are known and correctness must be demonstrated, ordinary computation remains superior to anything now called AI.

MODERN AI AS LEARNED MAPPING

Modern AI differs not because it ceases to compute, but because it shifts where the complexity resides.

In traditional software, the complexity is primarily in the rules written by the developer. In modern AI, much of that complexity is moved into a model that has been trained on data. The developer no longer specifies every useful path from input to output in explicit detail. Instead, an architecture is defined, data is selected, and optimization methods adjust the system until useful behavior emerges.

That is the crucial move.

A classical program follows instructions.

A modern AI system is shaped by examples.

The machine still performs calculations. But the behavior of the system now reflects learned parameters and statistical structure rather than a fully explicit chain of rules. What emerges is not a listed body of instructions, but a learned mapping from inputs to outputs.

This does not make the system mysterious. It makes it differently organized.

Rule-Based Systems and Learned Systems

The contrast between ordinary software and modern AI should not be drawn too crudely. There is no sharp boundary separating them, and many real systems are hybrids.

Historically, AI itself included rule-based systems — expert systems, symbolic reasoning engines, and search procedures. These were "AI" in the earlier technical sense, even though their behavior was largely explicit.

In current usage, the term most often refers to systems whose performance depends substantially on training, statistical modeling, and optimization.

Some systems are mostly explicit.
Some are mostly learned.
Many combine both.

A fraud detection system may combine hard-coded rules with a learned anomaly model. A medical system may combine statistical classification with fixed thresholds and human review. A language model may be paired with retrieval tools, filters, and rule-based constraints.

AI is not a separate kingdom. It is a mode of system design within computation.

Determinism and Probability

At the hardware level, computers remain deterministic machines. Instructions execute. Memory states change. Even large language models run on systems that are, within their operating limits, exact.

When modern AI is described as probabilistic, the meaning is different.

The system is organized around likelihoods — probabilities, confidence scores, weighted associations, and statistical estimates. A classifier assigns probabilities across categories. A recommender estimates likely preferences. A language model selects likely continuations.

Even when configured to produce stable outputs, the underlying process is not deductive procedure but statistical estimation.

This is one of the most important clarifications in the book:

> Traditional software derives results from rules.
>
> Modern AI estimates results from patterns.

The power comes from flexibility.

The weakness comes from the fact that likelihood is not the same as truth.

Exactness, Approximation, and Confidence

A calculator produces a precise result. A database returns a definite answer. A spreadsheet sums exactly within its limits.

AI systems operate differently.

They produce outputs that are approximate, ranked, or probabilistically selected. A system may identify a pattern, generate language, or recommend a course of action without certainty. Its value lies in usefulness under conditions where explicit rule-writing is impractical.

But this introduces a new kind of error.

A system may produce an answer that is fluent, plausible, and wrong.

This is not a malfunction in the ordinary sense. It is a consequence of the system's design. It is doing exactly what it was built to do — estimate and continue patterns.

This implications can be stated plainly:

> Classical systems fail through error in logic or calculation.
>
> AI systems often fail through misalignment of probability and truth.

Why the Distinction Matters

This is not a technical subtlety. It governs how these systems should be understood and used.

If a probabilistic system is treated as a deterministic one, it will be trusted too much. If a learned pattern system is mistaken for a thinking system, it will be misunderstood entirely.

Ordinary computation excels where rules are known and exactness is required. Modern AI becomes powerful where the world is too complex to be fully specified in advance.

The difference is:

A computer executes explicit procedure.

AI systems execute learned structure.

Both are computation. They are not the same kind of system behavior.

Sources and Notes

1. Hennessy, J. L., & Patterson, D. A. (2017). *Computer Architecture: A Quantitative Approach* (6th ed.). Morgan Kaufmann.

   A strong grounding source for what computers do at the architectural and procedural level, especially useful in keeping the discussion of computation concrete rather than mystical.

2. Turing, A. M. (1950). "Computing Machinery and Intelligence." *Mind*, 59(236), 433-460.

   Still the classic entry point for the question of machine intelligence and the slippage that often occurs between computation, imitation, and thinking.

3. Russell, S., & Norvig, P. (2021). *Artificial Intelligence: A Modern Approach* (4th ed.). Pearson.
   The standard broad survey of AI methods, useful here for distinguishing the larger field from the narrower popular meaning now associated with learned systems.

4. Bishop, C. M. (2006). *Pattern Recognition and Machine Learning.* Springer.
   Particularly valuable for understanding learned systems in probabilistic terms rather than as fixed rule engines.

5. Goodfellow, I., Bengio, Y., & Courville, A. (2016). *Deep Learning.* MIT Press.
   Foundational for understanding how modern models acquire behavior through training rather than through exhaustive manual specification.

6. Mitchell, M. (2019). *Artificial Intelligence: A Guide for Thinking Humans.* Farrar, Straus and Giroux.
   Useful for the boundary between public impression and technical reality, especially where fluent output is mistaken for understanding.

7. Searle, J. R. (1980). "Minds, Brains, and Programs." *Behavioral and Brain Sciences*, 3(3), 417-424.
   A valuable bridge to the next chapters, especially in preserving the difference between symbol manipulation and genuine understanding.

Chapter 3:
What is Called Thinking?

The prior chapter distinguished ordinary computation from modern AI. That explains much of what machines do, but it does not answer the question most often smuggled into public discussion: if a machine produces flexible, useful, and sometimes startling output, does it therefore think?

The answer depends first on what is meant by thinking.

A useful starting point is to distinguish between the thing itself and what we call it. The question is not merely what thinking is in the abstract, but what is meant when we apply the word *thinking* to something. This framing matters because present AI discourse often confuses performance with possession. A system may display outward marks associated with thought without possessing thought in the fuller human sense.

This framing echoes a line of inquiry found in the philosophical tradition, most notably in the work of Martin Heidegger, who examined not only thinking itself, but the assumptions embedded in how the term is used.[3]

A Pragmatic Definition

Names do not merely describe. They carry assumptions. And when those assumptions go unexamined, they can sweep us forward without our full intellectual participation.

---

[3] *What is Called Thinking?*, trans. J. Glenn Gray (New York: Harper & Row, 1968). See also *Being and Time* for Heidegger's broader treatment of understanding, interpretation, and the conditions under which meaning is disclosed.

For the purposes of this book, **thinking** will mean the **recollection and ordered processing of memory within a point of view capable of judgment**.

This definition establishes the entry condition for thinking and is intentionally minimal. Judgment, as used here, arises from a point of view grounded in lived memory and capable of anticipating and evaluating experience, not merely selecting outputs from patterns. Systems that do not meet this condition may simulate aspects of thought, but do not yet qualify as thinking in the sense used here.

Once this threshold is met, the question expands — into degrees, forms, and depth of thinking — but that discussion is not our current task. For now, it is simpler: to determine whether the threshold is crossed at all.

This is a working definition, not a final theory of mind. It is meant to be practical enough to separate mere calculation from thought without wandering into needless abstraction.

Three elements are involved.

First, **recollection**. Something from prior experience, knowledge, or context must be brought forward. Without recollection, there is no continuity between past and present, and no basis for reflection.

Second, **processing**. Thought does not merely retrieve. It compares, organizes, associates, weighs, combines, and revises. It may move by logic, analogy, intuition, or imagination. Processing turns what is recalled into something usable.

Third, **judgment**. Thought does not simply churn symbols. It brings what is recalled and processed to bear on some question, circumstance, or end. It asks, implicitly or explicitly: what matters here? what follows from this? what is true? what should be done? Without judgment, there may still be motion of information, but not thinking in the sense intended here.

Thinking begins in memory, but it is not exhausted by storage.

Thinking involves processing, but it is not exhausted by processing.
Thinking reaches its proper form only when memory and processing are gathered into judgment.

Related historical and philosophical treatments of thinking and intellect are summarized in Appendix A.

Recollection Is Not Mere Storage

This is an important differentiation because computers store and retrieve information extremely well, and this fact tempts people to equate storage with memory and retrieval with recollection. But the two are not the same.

A hard drive stores. A database retrieves. A cache preserves state. None of these facts, by themselves, amount to recollection in the human sense. Recollection is not merely the presence of prior data. It is the bringing-forward of what is relevant into present significance.

Human recollection includes sequence, salience, context, and meaning. It includes the difference between what one knows, what one half-remembers, what one has suffered, what one has learned, and what now matters because of the present question. Recollection is therefore not passive storage. It is memory entering judgment.

This is why recollection gives thought its continuity. It allows a thinker to stand in relation to prior understanding, revise it, regret it, defend it, or deepen it. It is part of what makes thinking owned rather than merely executed.

Storage alone does not satisfy the threshold of thinking, as it lacks both ordered processing and judgment within a point of view.

PROCESSING, INTUITION, AND INTELLECT

Processing is the active side of thought. It is what orders, tests, and reworks what recollection supplies. In one case this may appear as formal reasoning. In another it may appear as analogy, intuition, or rapid discernment.

Thinking, therefore, should not be reduced to explicit logic alone. Much of human thought moves faster than stepwise demonstration. Intuition is often the rapid comparison of present circumstance against accumulated experience. It is not magic. It is swift, tacit processing.

Intellect, as described by Jacques Barzun, may be understood as disciplined shorthand of the mind — stored intelligence refined into habits, symbols, and chains of reasoning by which the right knowledge can be brought to bear quickly.[4] Intellect is not a separate faculty floating above thought. It is one of thought's most efficient expressions.

This also helps with another confusion now common in the age of generative systems: the production of articulate language is easy to mistake for thought because language is the public garment of thought. But garments can be borrowed. A polished argument is not yet a proof of thinking. A convincing answer is not yet a proof of judgment. A flow of words may display rhetoric, association, or pattern completion without disclosing whether meaning is actually held by the speaker.

That contrast is ancient. Long before digital machines, there were already warnings about argument detached from truth. The ability to produce the stronger-seeming argument is not the same thing as genuine thought. Thinking is ordered not merely toward verbal success, but toward judgment. A fuller treatment of this pattern appears in Appendix F.

---

[4] *The House of Intellect.* Barzun uses "intellect" to describe disciplined, culturally formed reasoning that condenses knowledge into usable form. The usage here follows that general sense without adopting his broader critique of modern intellectual life.

These forms of processing contribute to thinking only when they are brought to bear within a point of view capable of judgment.

Experience, Memory, and Point of View

Human thinking operates not only on information, but on lived experience. Memory, in this sense, is not merely stored data but the retained structure of prior experience as it has been encountered, evaluated, and carried forward.

From this emerges a point of view — a continuity from which meaning is not only processed but held. This allows for anticipation as well as recollection. A thinker does not merely respond to the present input but brings prior experience to bear and projects possible outcomes forward. Judgment, therefore, is not simply the selection of an answer, but the evaluation of meaning within this continuity of experience.

AI systems do not possess lived memory or experience. They operate on recorded accounts of human experience present in their training data. As a result, they do not recall, inhabit, or anticipate experience from within a point of view. Instead, they generate responses that reflect patterns of how humans describe and evaluate such experiences.

This difference is essential. A system may produce a response that mirrors human judgment, but it does not follow that the system itself is judging. It is producing the form of judgment without possessing the condition from which judgment arises.

An Illustration: The Sunset

Consider the simple case of observing a sunset.

A human observer does not merely register colors and shapes. The experience is situated within memory — prior sunsets, moments in life, associations of place, time, and meaning. The present moment is not isolated; it is encountered as part of a continuity. From this continuity, a response may arise: *this is beautiful.*

That judgment is not simply a selection among alternatives. It reflects a point of view formed through lived experience. It carries preference, context, and often anticipation — the desire to remain, to reflect, or to return to such a moment again. The meaning is not only processed; it is held.

A modern AI system may produce the same statement. Given an image or description of a sunset, it may generate the response that the scene is beautiful. It may do so reliably, and with variation, depending on its training. Yet the similarity in output conceals a fundamental difference.

The system does not recall a prior sunset. It does not experience the present one. It does not anticipate a future experience. It operates on recorded accounts of how humans describe such moments and produces a response consistent with those patterns.

The difference is therefore not in the form of the response, but in its origin. The human judgment arises from within a point of view shaped by lived experience. The AI response is generated from learned associations in data.

The two may appear identical in expression. They are not identical in kind.

Why Output Is Not Enough

This is where the matter becomes especially important for AI.

Modern AI systems can now generate language that is fluent, organized, and often highly useful. They can summarize, explain, translate, compare, draft, and revise. To the user, these outputs can appear very close to thought itself.

Output alone is insufficient evidence.

A parrot may repeat words. A sophist may win an argument. A computer may produce a correct answer. A language model may generate a persuasive paragraph. None of these facts, taken by themselves, settles whether thinking is present.

Turing famously redirected attention from inner speculation to outward performance, and this was a profound move.[5] But for the purposes of this book, behavioral resemblance is not enough to settle the question. Searle's later requirement for understanding remains useful here: the manipulation of symbols according to form is not the same thing as understanding what those symbols mean.[6] Syntax is not yet semantics. Procedure is not yet thought.

This book therefore adopts a stricter standard. We are not asking whether machines can imitate some outward marks of thought. Plainly, they can. We are asking whether those marks amount to thinking in the fuller sense defined above.

Why Current AI Does Not Think

Current AI systems store vast amounts of data externally and encode learned structure internally in parameters, weights, and model states. They process inputs with remarkable speed and flexibility. They can detect patterns, continue sequences, rank possibilities, and generate responses that are often impressively coherent.

But none of this, by itself, amounts to thinking.

---

[5] Alan Turing, "Computing Machinery and Intelligence," *Mind* 59, no. 236 (1950). Turing proposed that questions about machine thinking should be approached through observable performance rather than speculation about inner states, leading to what is now known as the *Turing Test.*

[6] John Searle, "Minds, Brains, and Programs," *Behavioral and Brain Sciences* 3, no. 3 (1980). Searle's "Chinese Room" argument distinguishes between the formal manipulation of symbols (syntax) and the understanding of their meaning (semantics), arguing that the former does not entail the latter. Appendix E shows, in practical form, the same distinction Searle drew in theory: symbols may be manipulated correctly without being understood.

A language model does not recollect experience under a lived point of view. It conditions on present input and learned statistical structure. It does not remember in the human sense. It does not stand within the meaning of what it says. It does not operate from a point of view capable of judgment, and therefore does not meet the threshold established here for thinking. It does not weigh consequence as consequence, though it may reproduce patterns associated with prudent advice.

Put bluntly, a modern AI system can process symbols and patterns at extraordinary scale without inhabiting their meaning.[7] A machine may simulate some outward features of thought without possessing thought.

It may assist thinking, pressure thinking, extend thinking, or imitate thinking — but that is not the same as thinking.

BORDERLINE CASES AND THE LIMITS OF THE DEFINITION

A definition earns its place not by handling the easy cases, but by remaining clear at the boundary. Present AI now presses that boundary more forcefully than earlier systems did. Some systems retain conversational state, retrieve prior interactions, call external tools, decompose tasks into steps, revise plans after failure, and pursue objectives across extended sequences of action. To many observers, such behaviors appear very close to thinking. They therefore deserve direct treatment. If they are ignored, the argument of this chapter may appear cleaner than the technology warrants. If they are overstated, performance will once again be confused with possession.

The first pressure point is memory. Contemporary systems can preserve context windows, maintain external memory stores, retrieve prior preferences, and reintroduce earlier material when it becomes useful. This is real continuity of operation. It is not

[7] Symbols in the context of AI systems includes speech, writing, gesture, tone, and expression.

yet recollection in the sense used here. The difference lies between stored prior material and lived memory. A system may recover an earlier statement without remembering it as something it underwent, learned, regretted, suffered, or valued. It can retrieve the record of an event without standing in relation to that event as part of a continuing life. Persistence of state, however sophisticated, is therefore not yet recollection. It is continuity of information, not continuity of experience.

The second pressure point is planning. Some systems can decompose a task, rank options, call tools, revise a sequence after error, and continue toward a goal across many steps. This unquestionably increases capability. It does not by itself satisfy the threshold of thinking. Planning, in the functional sense, can be achieved through optimization, search, feedback, and control. Judgment, as used in this book, is more demanding. It is not merely the selection of a path toward a stated objective. It is the weighing of what should count as the objective, what competing goods or meanings are in tension, and what consequence matters from within a point of view. A system may optimize for an end without knowing the end as an end. It may adjust a plan without understanding what the outcome means.

A third pressure point arises from tool use and self-correction. A system that consults a database, runs code, searches a document set, compares alternative answers, or revises its output after critique can appear to be reflecting. Yet these abilities also admit of a lower explanation. They may indicate a more capable architecture, better orchestration, and stronger adaptive behavior. They do not by themselves establish that meaning is held rather than processed. The machine may improve its answer without understanding why the better answer is better. It may correct error without experiencing error as error. It may maintain consistency without owning what is being made consistent. The difference is subtle in appearance and decisive in kind.

The same point applies to systems that seem to "change their minds." In human beings, changing one's mind can be evidence of reflection, reconsideration, and judgment under a continuing point of view. In present AI systems, the appearance may arise from something narrower: altered context, a different prompt, expanded retrieved material, a stronger tool result, a different optimization tradeoff, or the ordinary instability of probabilistic output. A revised answer is not yet a revised judgment. It may be only a reconditioned continuation.

This is why behavioral resemblance cannot settle the matter on its own. A system may become increasingly difficult to distinguish from a thinker in limited interactions and still leave the underlying question open. The outward signs matter; they are not irrelevant. But they are evidentiary, not conclusive. A richer imitation of thought remains an imitation unless the threshold conditions of lived memory, point of view, and judgment are met. The public marks of thinking can be approached from the outside. That does not establish that they are being inhabited from within.

Reasonable people may therefore disagree at the boundary. A purely behavioral account would set the threshold lower. A more biological account might set it elsewhere. Some may reject the differentiation altogether and treat all sufficiently complex cognition as one continuous field. This book deliberately adopts a stricter threshold definition for a practical reason. The cultural danger of the present moment lies less in underrating machine inwardness than in granting it by implication. We are far more likely, today, to anthropomorphize fluent adaptive systems prematurely than to deny proven inwardness that has actually been shown.

For that reason, the burden of proof remains high. A system should not be called a thinker merely because its outputs are impressive, its continuity of operation has become more elaborate, or its behavior has become difficult to predict in detail.

Those facts may establish greater capability. They do not yet establish a point of view from which meaning matters. They do not yet establish recollection as lived memory. They do not yet establish judgment as the system's own.

This conclusion should not be mistaken for a proof of eternal impossibility. It is a judgment about the systems presently under discussion. If one day a machine were credibly shown to possess recollection as lived memory, judgment as its own, and a point of view from which consequence genuinely mattered, then the question would have to be reopened. But that is not the condition of contemporary AI. What we have now is more powerful than older automation, more adaptive than traditional software, and more outwardly thought-like than earlier machines. It is not, on the evidence presently available, thinking in the sense defined here.

A Further Test: On Derivation and Origination

The prior section tested the boundary of thinking by examining systems that approach its outward form. A further clarification follows by stepping back from behavior to the source of expression itself. A further clarification may be made by stepping back from behavior and asking a more basic question: what is the source of the expression itself?

Human expression does not begin as data. It begins in experience. A person encounters the world, processes it as beauty, loss, meaning, or significance, and then expresses something of that experience in language, image, or form. What remains — literature, music, philosophy, history—is not the experience itself, but its residue.

Another human being may encounter that residue and, within his own experience, reconstitute something of what it meant. In this sense, human understanding is **bidirectional**. It moves from experience into expression, and from expression back into experience. Meaning is not merely transmitted. It is re-formed within a new point of view.

Artificial intelligence operates differently. It has access to the same residue — the accumulated record of human expression — but not to the experience from which that expression arose. Its operation is therefore **unidirectional**. It proceeds from recorded expression to new expression through learned pattern structure, without returning to the experiential ground from which meaning is formed.

A clarification is required at this point. Human beings do not operate only through reconstituted experience. They are also capable of working directly on symbols — rearranging language, manipulating number, and extending patterns without fully returning to the lived condition from which those symbols arose. In this narrower sense, human cognition can resemble the operation of artificial systems.

But the resemblance is incomplete. Human thought is not confined to symbolic manipulation. It can return from expression to experience, and from that experience form new judgment and meaning. Artificial intelligence operates only in the first mode. It processes patterns within expression, but does not reconstitute the experiential ground from which meaning arises.

This difference may be stated plainly. Artificial intelligence can derive. It can recombine, extend, and vary patterns within what has already been expressed. Its outputs may be novel, but that novelty remains bounded by the structure of prior expression.

Human beings can also derive. But they do not only derive. They can originate. Because human thought is grounded in lived experience, it can introduce new meaning-structures that are not reducible to prior patterns alone. It can reorganize how the world is seen, not merely extend how it has already been described.

The crucial difference is therefore not between repetition and novelty. Both humans and machines can produce novelty. It lies deeper.

Artificial intelligence produces novelty within a learned space. Human thought can alter the space in which meaning itself is formed.

This is not a separate definition of thinking, but a further test of it. A system that does not originate meaning, but only extends prior expression, has not crossed the threshold established in this chapter. It operates with remarkable power within the record of human thought, but it does not stand within the condition from which that record was produced.

Why This Distinction Matters

This conclusion matters not because it diminishes AI, but because it clarifies it.

If we wrongly attribute thinking to machines, we are likely to misunderstand both their strengths and their dangers. We will trust fluent output too readily. We will mythologize systems that are better understood as learned pattern engines. And we may overlook the fact that much of the real risk of AI lies not in machine thought, but in human use of non-thinking systems at enormous scale.

At the same time, to say that current AI does not think is not to say that it is trivial, useless, or unintelligent. Thinking and intelligence are not identical. A system may adapt effectively without reflective thought. That is the next question.

We therefore leave thinking here, with one conclusion stated as plainly as possible:

Current AI does not think.

It computes, models, predicts, ranks, generates, and adapts. Those are remarkable capabilities. They are not thought.

The difference is not in the form of the output, but in the condition from which it arises.

Sources and Notes

1. Heidegger, M. (1968). *What Is Called Thinking?* Harper & Row.
   Direct philosophical inspiration for the title and for the distinction between what thinking is and what is meant when something is called thinking.
2. Searle, J. R. (1980). "Minds, Brains, and Programs." *Behavioral and Brain Sciences*, 3(3), 417-424.
   Essential for preserving the distinction between symbol manipulation and understanding.
3. Dreyfus, H. L. (1972). *What Computers Can't Do: A Critique of Artificial Reason.* Harper & Row.
   A strong critique of attempts to reduce human thought to formal procedure alone.
4. Barzun, J. (1959). *The House of Intellect.* Harper & Row.
   Especially useful for the disciplined notion of intellect as refined and rapid reasoning built from stored intelligence.
5. Turing, A. M. (1950). "Computing Machinery and Intelligence." *Mind*, 59(236), 433-460.
   Still the classic operational treatment of machine intelligence, and useful here partly because this book adopts a stricter distinction than behavioral imitation alone.
6. Nagel, T. (1974). "What Is It Like to Be a Bat?" *The Philosophical Review*, 83(4), 435-450.
   Helpful for the question of subjective point of view, which remains central to the difference between output and thought.

Chapter 3: What is Called Thinking?

7. Aristophanes (423 BC). *The Clouds.*
   Not a technical source, but an early and still valuable warning about argument detached from truth.

Chapter 4:
What is Called Intelligence?

With thinking now more clearly defined, the question of intelligence becomes easier to handle. Much of the confusion in present AI discourse comes from treating thinking, intelligence, intellect, consciousness, and judgment as interchangeable. They are not.

The conclusion of the prior chapter was that current AI does not think. A common mistake is then made in the opposite direction: if it does not think, it must not be intelligent. That conclusion does not follow. Thinking and intelligence are related, but they are not the same thing. A system may be intelligent without possessing a lived point of view, and it may adapt effectively without reflective thought.

That distinction is essential to this book, because much of what is impressive, useful, and potentially dangerous about modern AI depends precisely on the fact that it can be intelligent in an important sense while not thinking in the fuller sense established earlier.

Intelligence as Adaptability

For the purposes of this book, intelligence will mean adaptability.

More precisely, intelligence will mean the capacity of a system to **produce effective adjustment under varying conditions** relative to some task, environment, or objective. A system is intelligent to the degree that it responds effectively when inputs, context, or demands change, without requiring every case to be specified in advance.

This is a deliberately functional definition. It does not assume consciousness. It does not assume inner life. It does not assume morality, self-awareness, or reflective judgment. It asks a narrower question: when conditions vary, can the system adjust in a way that remains effective?

This definition has two advantages. First, it can be applied across domains without forcing everything into an exclusively human mold. Second, it allows intelligence to be discussed in degrees and kinds rather than as an all-or-nothing possession. Human intelligence, animal intelligence, and machine intelligence need not be identical in form to be meaningfully discussed under one heading.

Adaptation Is More Than Fixed Reaction

A qualification is needed immediately, because not every response to a changing condition deserves to be called intelligence.

A fuse blows when current rises too high. A thermostat switches on when temperature drops below a set point. A motion sensor activates when movement crosses a threshold. These are useful control mechanisms, but they are too narrow and too fixed to bear much weight as intelligence in the sense intended here.

These systems produce reactions, not adaptation. They do not bring a broad range of prior structure to bear on new conditions, nor do they generalize meaningfully across contexts. Each operates within a limited loop, executing a fixed response to a defined signal.

By contrast, certain biological systems exhibit a different kind of behavior. A fungal network, for example, does not respond through a single fixed rule. It grows, retracts, redistributes resources, and adjusts its structure in response to varying environmental conditions such as nutrient availability, obstacles, and competing organisms. The response is not identical from case to case, nor is it confined to a single signal-response loop.

Such systems approach what may be considered the lower boundary of intelligence. They exhibit adaptive behavior across conditions, but in a limited and highly constrained form. Whether such systems are properly described as intelligent is less important than the contrast they reveal: they are no longer purely reactive, yet they do not demonstrate the broader capacity for learning, abstraction, or transfer that characterizes more developed forms of intelligence.

Intelligence begins in earnest where a system can absorb structure from prior cases and use it effectively in new ones.

This is why learning is such a powerful marker of intelligence, though not its only marker. Systems that exhibit deeper adaptability do not merely adjust — they improve their ability to adjust. They incorporate prior structure and apply it across new conditions.

Where such learning is present, adaptability becomes self-improving, and the case for intelligence becomes clear. Where it is absent, what appears to be adaptability often reduces to fixed or narrowly bounded response.

Degrees and Kinds of Intelligence

Intelligence should be understood as admitting degree and kind.

A child learning language, a hawk tracking prey, a chess engine evaluating positions, and a modern AI model answering unfamiliar prompts are not intelligent in the same way. Their range, depth, embodiment, transfer, and context differ greatly. But all can be discussed under the heading of intelligence insofar as they exhibit adaptive behavior that is not exhausted by a single fixed response.

Human intelligence is distinctive. It is deeply bound up with language, social life, culture, imagination, and reflective judgment. It can transfer across remarkably different domains, revise itself in light of principle, and bring not only skill but meaning to bear on circumstance.

Animal intelligence is often more situated and embodied, yet can be extraordinarily effective within the conditions of life in which it operates.

Machine intelligence, in its present form, is different again. It is often narrow, uneven, data-dependent, and task-bounded. It may be formidable within a defined domain and brittle outside it. But difference in kind does not eliminate the underlying family resemblance. Intelligence need not look human in order to be real.

Why Current AI Qualifies as Intelligent

Current AI qualifies as intelligent under this functional definition because it does more than repeat one explicit routine. Modern systems are trained on data, internalize statistical structure, and then use that learned structure to respond adaptively to new cases.

A language model can answer unfamiliar questions, summarize documents it has not seen before, shift tone and format to match context, translate across languages, classify text, and produce varied outputs from varied prompts. A vision model can identify patterns in previously unseen images, detect anomalies, or distinguish categories under noisy conditions. Recommendation systems can adjust rankings as user behavior changes. Control systems informed by machine learning can adapt performance across environments more flexibly than older hand-authored systems alone.

None of this requires thinking in the sense already established. But it does require adaptability. The system is not merely replaying a fixed rule chain for every case. It is applying learned structure to variation. That is why the word *intelligence* is not misplaced here, provided it is used carefully.

To deny the term altogether would create more confusion, not less. It would force us either to pretend that clearly adaptive systems are not adaptive, or to reserve the word *intelligence* only for human-style cognition. Neither move is helpful. The more disciplined approach is to say plainly that present AI is intelligent in a real but limited sense.

Intelligence Is Not Thinking

That statement now becomes sharp.

Thinking, as defined in the prior chapter, requires the recollection, ordering, and processing of memory within a point of view capable of judgment. Intelligence does not require that full threshold. A system may adapt effectively without possessing lived memory, without holding meaning within a point of view, and without judging as a thinker does.

Put plainly:

> Thinking requires a point of view capable of judgment.
> Intelligence requires effective adaptation.
> The first is absent in current AI.
> The second is plainly present.

This is the point on which much current AI discussion goes astray. People see adaptive output and infer thought. But adaptability is not yet thinking. Even where adaptability is highly developed — where a system learns and improves its responses across conditions — this remains distinct from thinking, which requires judgment from within a point of view.

An AI system may successfully classify a medical image, recommend a course of action, or generate a remarkably useful explanation. That is evidence of intelligence in the adaptive sense used here. It is not evidence that the system experiences, remembers, anticipates, or judges from within a lived continuity of meaning.

Intellect Is Not the Same Thing Either

A further separation is useful here. Intelligence should not be confused with *intellect.*

Intellect, in the sense already described, is not mere adaptability. It is intelligence refined, stored, disciplined, and made rapidly available through symbols, habits of reasoning, and shared forms of culture. Intellect belongs not only to the individual mind but to the accumulated life of civilization. It is intelligence brought into ordered, communicable, and often high-speed use.

This matters because some AI outputs can appear highly intellectual. They can mimic learned discourse, summarize arguments, and move through chains of reasoning with impressive fluency. But fluency in the forms of intellect is not the same as possessing intellect. Present AI operates on the recorded products of human intellectual life. It does not stand within that life as a participant who inherits, judges, and contributes from within experience.

This book therefore treats thinking, intelligence, and intellect as related but distinct:

> **Thinking** concerns judgment within a point of view.
> **Intelligence** concerns adaptability.
> **Intellect** concerns disciplined and culturally formed and stored intelligence brought to bear with speed and structure.

The differences are not ornamental. They prevent category errors.

WHY THE DISTINCTION MATTERS

This chapter is not merely a definition. It is setting a boundary that later chapters will depend upon.

If we deny intelligence to AI altogether, we will underestimate its capability and misuse its limits. If we attribute thinking to it where thinking is absent, we will mythologize it and inflate fears that do not correspond to present reality. The more accurate position is the more useful one:

> Current AI is intelligent in the sense that it adapts effectively within domains.
>
> Current AI does not think in the sense that it does not judge from within a point of view.
>
> And current AI does not possess intellect in the fuller human and civilizational sense.

This is not verbal housekeeping. It is the basis for sober analysis. Systems can be powerful, consequential, and even dangerous without being thinkers. Indeed, part of the difficulty of modern AI is that it combines real adaptive capability with the absence of understanding, judgment, and lived stake. That combination will matter greatly when we turn later to society and fear.

This adaptive capability is not merely reactive. Modern AI systems improve their ability to respond by incorporating structure from prior cases and applying it across new situations. In this sense, their adaptability is not fixed, but learned. This capacity for learning — for improving how they adapt — marks a deeper form of adaptability and strengthens the case for describing such systems as intelligent.

CONCLUSION

For the purposes of this book, intelligence is best understood as adaptability. That definition allows us to recognize what current AI genuinely does without granting it more than it has earned.

Modern AI systems do not think. They do not inhabit meaning, possess lived memory, or judge from within a point of view. But they do adapt. They generalize, rank, classify, generate, and respond under changing conditions in ways that are often highly effective. In that functional and important sense, they are intelligent.

This conclusion is foundational for what follows. It allows the book to avoid both naïve dismissal and exaggeration. AI is neither a mere calculator nor a thinking mind. It is an adaptive artificial system — powerful, limited, and in need of clear description before it can be judged soberly.

Intelligence, as used here, is real. It is not the same as thought.

Sources and Notes

1. Legg, S., & Hutter, M. (2007). "A Collection of Definitions of Intelligence."
   Useful as a survey of how widely the concept of intelligence has been defined, and especially helpful in supporting a functional rather than purely anthropocentric approach.

2. Chollet, F. (2019). "On the Measure of Intelligence."
   Valuable for emphasizing generalization, adaptation, and skill acquisition across conditions rather than narrow benchmark performance alone.

3. Gardner, H. (1983). *Frames of Mind: The Theory of Multiple Intelligences.* Basic Books.
   Important for broadening the discussion beyond a single narrow model of intelligence.

4. Sternberg, R. J. (1985). *Beyond IQ: A Triarchic Theory of Human Intelligence.* Cambridge University Press.
   Helpful in reinforcing that intelligence includes practical and adaptive dimensions, not only formal test performance.

5. Turing, A. M. (1950). "Computing Machinery and Intelligence." *Mind*, 59(236), 433-460.
   A classic reference for machine intelligence, especially useful as a contrast with the stricter distinctions used in this book.

6. Russell, S., & Norvig, P. (2021). *Artificial Intelligence: A Modern Approach* (4th ed.). Pearson.
   Standard technical grounding for AI methods and capabilities across a broad range of domains.

7. Barzun, J. (1959). *The House of Intellect.* Harper & Row.
   Useful here not for intelligence in general, but for distinguishing intellect from both intelligence and thinking.

Chapter 5:
Computing Power versus Programming Power

If current AI can be intelligent without thinking, the next question is how such capability became possible.

It did not appear as a single invention. Artificial intelligence did not arrive — it emerged through long interaction.

That accumulation was driven by two interacting forces: **computing power** and **programming power**. The first concerns what machines can process, store, move, and execute. The second concerns how effectively logic can be organized so that those machines perform useful work. Neither is sufficient alone. Fast hardware with weak methods remains crude. Elegant methods without adequate hardware remain trapped by physical limits. Modern AI emerged when both trajectories rose together and began reinforcing one another.

Later chapters will examine that interaction historically and quantitatively. Here the task is simpler: to define the two terms clearly and explain why their differentiation matters.

WHAT IS MEANT BY COMPUTING POWER

In this book, *computing power* means more than processor speed. It includes the effective capacity of a machine or distributed system to perform operations, store data, move data, and do so at useful scale and speed. Processing throughput, memory, storage, bandwidth, parallelism, and specialized hardware all belong here.

This broader definition matters because AI does not depend on one component alone. A system may have fast arithmetic yet still be constrained by memory, data movement, latency, or storage. In modern AI, such bottlenecks are often as important as raw clock speed. A model cannot be trained effectively if the data cannot be fed fast enough, if memory cannot hold the necessary state, or if communication across processors becomes the limiting factor. What matters in practice is the total effective capacity of the computational arrangement.

Computing power therefore refers to the machine-side ability to bear computational load.

As that load-bearing capacity increases, more becomes possible. Larger data sets can be processed. More parameters can be trained. More experiments can be run. More complex models can be deployed in practical time. Some AI methods were understood in simplified or theoretical form long before they became useful at scale. What changed was not always the underlying idea, but the machine environment that finally made the idea operational.

What is Meant by Programming Power

Programming power is the less obvious term, but in many ways the more important one.

By programming power, this book does not mean only the talent or intelligence of an individual programmer. It means the effective power of our methods for turning intention into computational behavior. Algorithms, data structures, programming languages, abstractions, software architectures, libraries, frameworks, optimization methods, debugging tools, testing discipline, reusable code, collaborative development practices, and machine-assisted development all contribute to it.

Programming power rises when the same human effort can produce systems that are more capable, more reliable, more scalable, or more sophisticated than before. It rises again when prior work becomes reusable, composable, and widely available to others.

This is one of the great cumulative features of software civilization. A high-level language can compress what once required vast amounts of low-level instruction. A mature library can preserve years of hard-won expertise and make it available in a few lines of code. A framework can convert what was once research-level difficulty into ordinary engineering practice. An optimization method can turn infeasible training into routine procedure. In this sense, programming power is not just coding. It is accumulated method.

If computing power is the machine's capacity to carry load, programming power is our capacity to make that load productive.

Why the Distinction is Necessary

Discussions of AI often drift toward one-sided explanations.

One school attributes progress almost entirely to hardware: more compute — that is, greater effective processing capacity — better chips, larger clusters, faster interconnects. Another attributes it largely to algorithms: better architectures, cleaner abstractions, improved optimization, better tooling, smarter methods. Both observations capture part of the truth. Neither captures the whole.

A more powerful machine does not automatically produce a more capable system if the methods are poor. A weak algorithm run at enormous scale can still waste resources. Conversely, a better method may show only limited value if the available computing environment is too small, too slow, or too constrained to let the method unfold.

This is why many important ideas in AI appear twice in history. They appear once as theory, often before the surrounding conditions are ready, and then again later as practical breakthroughs once the hardware, software, data, and engineering ecosystem catch up. What looked premature in one era becomes obvious in another, not because the idea changed completely, but because the surrounding trajectories finally converged.

The use of computing power and programming power helps prevent simplistic explanations of AI progress. It makes clear that AI capability is not the gift of hardware alone, nor the triumph of clever code alone. It is the product of their convergence.

Mutual Reinforcement

The relationship between computing power and programming power is not merely additive. It is reciprocal.

Limited hardware pressures programmers toward efficiency, compact representation, clever abstraction, and disciplined use of resources. Better hardware permits larger models, deeper experimentation, richer representations, more ambitious software systems, and wider deployment. Those improved methods then increase the value of further hardware advances by revealing new opportunities and new bottlenecks. Better software makes better hardware more worthwhile; better hardware makes better software more possible.

This reciprocal pressure is one of the main engines of computing history.

When hardware is scarce, programming practice becomes severe and economical. When hardware expands, programming ambition expands with it. When programming methods improve, they often demand more of hardware in memory, throughput, or parallel execution. Each side pushes on the other.

Over time, this becomes a reinforcing cycle:

- more computing power enlarges the space of feasible methods,
- better methods enlarge the value of additional computing power,
- and the combination raises the ceiling of what can be built.

Modern AI is one of the clearest expressions of that cycle.

From Ordinary Computation to Learned Systems

This interaction matters especially for AI because learned systems are hungry in both dimensions.

A classical procedural system may be difficult to write, but once written it often executes within relatively stable requirements. A learned system is different. It requires adequate computing power to train on large bodies of data, to optimize parameters, to evaluate performance, and then to perform inference at useful speed. At the same time, it requires programming power to define model architectures, data pipelines, objective functions, optimization strategies, training procedures, evaluation methods, and deployment infrastructure.

A modern language model is not simply "a lot of compute". Nor is it simply "good software". It is the convergence of enormous computational capacity with mature algorithms, scalable architectures, robust frameworks, disciplined engineering practice, and increasingly reusable software ecosystems.

Remove either side and the system collapses.

Without sufficient computing power, the methods cannot scale. Without sufficient programming power, the hardware has no useful intelligence to express.

This is why the two-driver framework is so important. It explains not only why AI became possible, but why its progress has been uneven. At some times, hardware surged ahead and methods lagged. At other times, programming innovation extracted new capability from relatively stable hardware. At still other times, both rose together and produced visible inflection points. The modern era of AI reflects repeated episodes of such convergence.

Programming Power as Stored Method

One further point is worth stating plainly.

Programming power is not just what an engineer can write from scratch today. It is also what civilization has already learned to encode and hand forward.

Every mature programming language, every reusable library, every training framework, every standard abstraction, every debug tool, every open-source repository, and every body of engineering practice stores prior intellectual labor. What once required inventing can later be imported. What once demanded specialists can later be applied by general practitioners. What once consumed months of work can later be assembled in hours.

This is not incidental. It is one of the major reasons programming power compounds over time.

The significance for AI is enormous. As frameworks, pre-trained models, optimization methods, and deployment tooling mature, they reduce the distance between an idea and a functioning system. Programming power therefore grows not only by making code better, but by making capability more accessible. It compresses prior effort into present leverage.

Why This Chapter Matters

The distinction between computing power and programming power is not an abstract taxonomy. It is the conceptual basis for the book's later analysis.

If AI is viewed only as software, its dependence on physical computational conditions is obscured.

If AI is viewed only as hardware, the role of accumulated method disappears.

If AI is treated as a sudden event, the long accumulation that made it possible is misunderstood.

The more accurate picture is this:

> Artificial intelligence emerged through the long interaction of machine capacity and method capacity.
>
> Computing power increased what could be carried.
>
> Programming power increased what could be made useful.
>
> Their convergence produced the adaptive artificial systems now called AI.

This framework also helps explain why the field does not move in a straight line. Progress is not governed by a single curve. It is governed by the shifting interaction of at least these two major forces, each of which can accelerate, plateau, or redirect the other.

In the quantitative chapters that follow, these two drivers are not treated impressionistically, but as normalized historical proxy indices constructed from explicit component scoring and milestone anchors with the underlying construction detailed in Appendix C (Tables C1–C5). The aim is not to claim exact annual measurement, but to represent the changing interaction of machine-side and method-side capacity in a disciplined first-order form.

CONCLUSION

Artificial intelligence did not appear fully formed. It developed through the mutual reinforcement of computing power and programming power.

Computing power provided the capacity to process, store, move, and execute at scale.

Programming power provided the ability to convert that capacity into useful, adaptive behavior through algorithms, abstractions, frameworks, and engineering method.

Neither alone explains modern AI.

Together, they explain why systems once theoretical became practical, why narrow tools became widely adaptive, and why progress arrived not as magic, but as convergence.

This is important for what follows. Once the two drivers are separated conceptually, we can attempt a more disciplined assessment of AI capability over time and better understand the major changes in its rate of development.

Sources and Notes

1. Hennessy, J. L., & Patterson, D. A. (2017). *Computer Architecture: A Quantitative Approach* (6th ed.). Morgan Kaufmann.
   Foundational for the machine-side realities of computation: throughput, memory, architecture, and the practical limits that shape what software can do.

2. Brooks, F. P. (1995). *The Mythical Man-Month* (Anniversary ed.). Addison-Wesley.
   Valuable not for AI specifically, but for the enduring realities of software engineering, complexity, and the difference between raw effort and effective method.

3. Amdahl, G. M. (1967). "Validity of the Single Processor Approach to Achieving Large-Scale Computing Capabilities."
   A classic reference for the limits of parallel gains and a useful reminder that more hardware does not translate mechanically into unlimited performance.

4. Sutter, H. (2005). "The Free Lunch Is Over."
   Helpful for the historical shift from easy gains in raw processor speed toward concurrency, parallelism, and more demanding software design.

5. Russell, S., & Norvig, P. (2021). *Artificial Intelligence: A Modern Approach* (4th ed.). Pearson.
   Useful for placing modern AI methods in the broader history of the field and for avoiding overly narrow explanations of progress.

6. Goodfellow, I., Bengio, Y., & Courville, A. (2016). *Deep Learning.* MIT Press.
   Important for understanding how model architecture, optimization, training practice, and computational scale work together in modern AI.

Chapter 6:
The AI Capability Index

With the two principal drivers now separated, one further step becomes necessary. If computing power and programming power remain only descriptive ideas, discussion of AI progress quickly drifts into impression, slogan, or hype. Terms such as "more compute" and "better software" may be directionally true, but they do not yet provide a disciplined way to compare eras, identify acceleration, or test whether a claimed advance is structural or merely visible.

For that reason, this book introduces an AI Capability Index.

The purpose of the index is not to pretend that all of artificial intelligence can be reduced to a single number. It cannot. The purpose is narrower and more practical: to provide a structured approximation of the historical capacity to build and deploy adaptive artificial systems. The index is therefore an analytical tool, not a metaphysical claim. It measures capability, not thought, and certainly not wisdom.

WHY AN INDEX IS NEEDED

A purely narrative account of AI development has two weaknesses.

First, it tends to over-credit visible breakthroughs while under-credit the long accumulation that made them possible. A model appears, captures public attention, and is treated as though it sprang into existence on its own. The decades of underlying buildup are then obscured.

Second, a narrative account tends to flatten differences between types of progress. A new chip, a better optimization method, a reusable framework, a training breakthrough, and a polished product launch may all be described loosely as "AI progress," even though they operate at different levels. Without a framework, one cannot tell whether the field has advanced in raw machine capacity, in method, in accessibility, or merely in presentation.

An index does not remove judgment — it imposes discipline.

Capability as Interaction, Not Addition

The central claim of this chapter is simple:

> AI capability is best approximated as the interaction of computing power and programming power.

In simplified form:

> AI Capability Index $\approx$ CPI $\times$ PPI,
> where CPI = Computing Power Index,
> and PPI = Programming Power Index.

This should not be read as a law of nature. It is a working model. Its value lies in the structure of the relationship it expresses.

The product form matters. If the two drivers were merely added, one could too easily compensate for weakness in one dimension by strength in the other. But that is not how modern AI actually develops. Extraordinary hardware with crude methods remains limited. Elegant methods with insufficient hardware remain unrealized. The product captures this mutual dependence. If either factor is weak, overall capability remains constrained. When both rise together, capability compounds.

The product form is not intended to represent a precise law of AI development. More complex formulations — involving powers, weights, or nonlinear scaling — could be constructed. However, such forms would introduce assumptions about the relative importance of each driver that are not justified at this level of analysis. The method is therefore not chosen for precise equivalence to the real world, but for discipline. It is the simplest expression that captures the essential structural relationship: mutual dependence.

This is precisely what Chapter 5 argued in conceptual form. It is a first-order model, sufficient to capture the dominant interaction without over-specifying it. The index now expresses that same argument quantitatively, providing an analytical representation of what we know while minimizing the risk of claiming a level of precision that the model cannot support.

What the Two Component Indices Represent

The **Computing Power Index** represents the machine-side capacity to carry computational load. It includes not only raw processing throughput, but also memory, storage, bandwidth, parallelism, and specialized hardware relevant to AI workloads. What matters is the effective capacity of the computational system, not a single headline specification.

The **Programming Power Index** represents the method-side ability to turn computational capacity into useful adaptive behavior. It includes algorithms, abstractions, frameworks, optimization methods, software architecture, reusable libraries, tooling, and the broader engineering practices that compress prior intellectual labor into present leverage.

Each of these is itself a composite idea. The aim is not to claim that one specification or one programming language explains an era, but to capture the general rise of each driver in a way that can be compared over time.

A Note on Data

An obvious objection arises here: what about data?

The objection is well taken. Modern AI depends heavily on large quantities of usable data. However, data does not act as a first-order driver. In many cases, data exists in abundance. The limiting factor is not its presence, but its usability.

That usability is determined by computing power and programming power — the capacity to process, structure, and convert data into learned capability. Data without sufficient computing power cannot be processed at scale. Data without sufficient programming power cannot be organized, modeled, or transformed into effective systems.

In this sense, data is best understood as a necessary substrate rather than a primary driver. Its effect enters the system through storage, bandwidth, tooling, model design, training methods, and the broader engineering environment that makes it usable.

This does not make data unimportant. It means only that, for the purposes of a disciplined first-order model, the principal drivers remain computing power and programming power, while data serves as the material upon which those drivers operate.

How the Annual Series Are Constructed

The detailed derivation appears in Appendix C, but the governing structure should be stated here.

Each index is built as an annual proxy series from 1920 through 2024. For both computing power and programming power, the model uses explicit component categories scored on a common scale. Where meaningful quantitative proxy data are available, those proxies are mapped into the component scores. Where they are not, milestone anchors are assigned and annual values interpolated between them. The weighted component scores are then aggregated into raw indices and normalized to a 0–1 scale.

The AI Capability Index is the product of the resulting normalized CPI and PPI series.

This is therefore not a black-box statistical inference model. It is a proxy-based historical scoring model whose assumptions are visible and explicit in structure. Tables C1–C5 in Appendix C provide the annual values that were used in constructing the curves presented in this book.

NORMALIZATION AND HISTORICAL USE

Because the book compares long stretches of time rather than a single generation of hardware or software, each component index is normalized. The exact value assigned to any single year is less important than the broad shape of the resulting curve — its slope, its rate of change, and the periods in which the two drivers begin to reinforce one another more strongly.

This also means that the index should not be read as a scoreboard. It is a historical instrument.

Its purpose is not to claim that 1997 was "exactly" this capable or that 2017 was "precisely" that capable. Its purpose is to represent the changing capability envelope of different eras in a disciplined and reproducible form.

WHAT THE INDEX MEASURES — AND WHAT IT DOES NOT

The index measures the evolving capacity of a technological civilization to produce adaptive artificial systems. It is best understood as an approximation of the practical capability envelope of a given era.

It does not measure:

- thought
- consciousness
- judgment
- wisdom
- moral status
- or social desirability

Nor does it directly measure whether a system is used well, safely, or responsibly. A civilization may possess enormous AI capability and still deploy it poorly. Capability and judgment are not the same thing.

This distinction is essential. The index belongs to the technological side of the book, not the philosophical side. It helps explain how much adaptive artificial capability could be built. It does not settle what that capability means, nor whether it should be trusted without restraint.

Why Learning Sits Near the Center of the Index

Chapters 3 and 4 established that current AI does not think but is intelligent in a limited sense. The present chapter adds a further clarification: the capability worth measuring is not merely the ability of a system to respond, but the historical ability to build systems whose adaptability — intelligence, in the sense defined earlier — can be extended and improved over time.

This is why the index is framed around the interaction of machine capacity and method capacity. Learning at useful scale requires both. It requires enough computational force to process large bodies of data and enough programming sophistication to define models, objectives, optimizers, pipelines, and evaluation methods through which prior structure can be absorbed and applied to new conditions.

In this sense, the index does not measure "AI" as a fashionable label. It measures the accumulated capacity to build systems whose adaptability is no longer fixed, but learned.

Why the Index Matters

Once this framework is in place, later chapters can ask better questions.

Instead of asking vaguely whether AI "suddenly arrived," we can ask whether one or both of the core drivers experienced a sustained change in rate.

Instead of treating every celebrated model as an isolated breakthrough, we can ask whether it was enabled by a broader rise in capability already underway.

Instead of assuming that progress must be linear, we can look for periods in which the two drivers converged strongly enough to produce a markedly steeper increase in practical capability.

This is the analytical value of the index. It allows the book to move from impression to structure.

Conclusion

The AI Capability Index is a disciplined approximation of the historical capacity to build adaptive artificial systems. It rests on a simple but powerful claim: AI capability emerges not from hardware alone or software alone, but from the interaction of computing power and programming power.

The product form reflects that dependence. If either driver remains weak, capability is constrained. When both rise together, practical AI capability compounds.

The full derivation appears in Appendix C. What matters here is the governing logic. Once the two drivers are defined and their interaction expressed quantitatively, the development of AI can be examined more soberly — not as myth, not as magic, but as accumulation, convergence, and measurable change over time.

The index developed here provides the historical instrument used in the next chapter to examine how artificial intelligence developed over time.

SOURCES AND NOTES

1. Moore, G. E. (1965). "Cramming More Components onto Integrated Circuits."
   A foundational marker for the long arc of increasing computational density and machine capacity.

2. Hennessy, J. L., & Patterson, D. A. (2017). *Computer Architecture: A Quantitative Approach* (6th ed.). Morgan Kaufmann.
   Essential for understanding why effective computing power includes more than processor speed alone.

3. Goodfellow, I., Bengio, Y., & Courville, A. (2016). *Deep Learning.* MIT Press.
   Useful for the interaction between model design, optimization, and computational scale in modern AI.

4. Russell, S., & Norvig, P. (2021). *Artificial Intelligence: A Modern Approach* (4th ed.). Pearson.
   A broad grounding source for AI methods and their historical development.

5. Chollet, F. (2019). "On the Measure of Intelligence."
   Helpful for keeping capability, adaptation, and generalization conceptually distinct from raw benchmark performance.

Chapter 7: Observations of AI Development Over Time

With the Computing Power Index, the Programming Power Index, and their product now defined, the curves may be read as a single historical pattern. Their value does not lie chiefly in exact annual values, but in **broad changes of slope**.

The curves are smoother than the public narrative of AI. That is not a defect of the model; it is one of its findings. The curve is continuous, even though the public experience of it is not.

Artificial intelligence did not rise as a sequence of miracles. It accumulated over long periods, then became newly visible when the two principal drivers converged strongly enough to make new kinds of systems practical.

The years identified below should therefore be read as turning periods, not isolated instants. The aim is not to force history into artificial boundaries, but to identify where the combined capability envelope changed character.

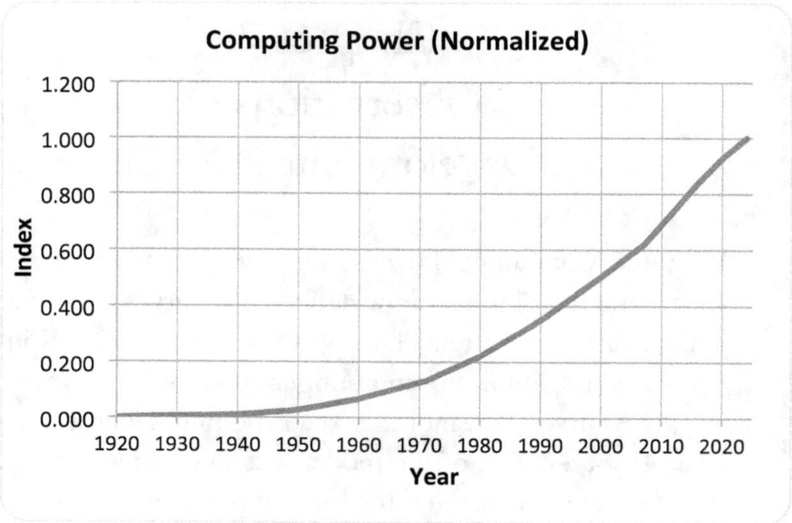

Computing Power (Normalized)

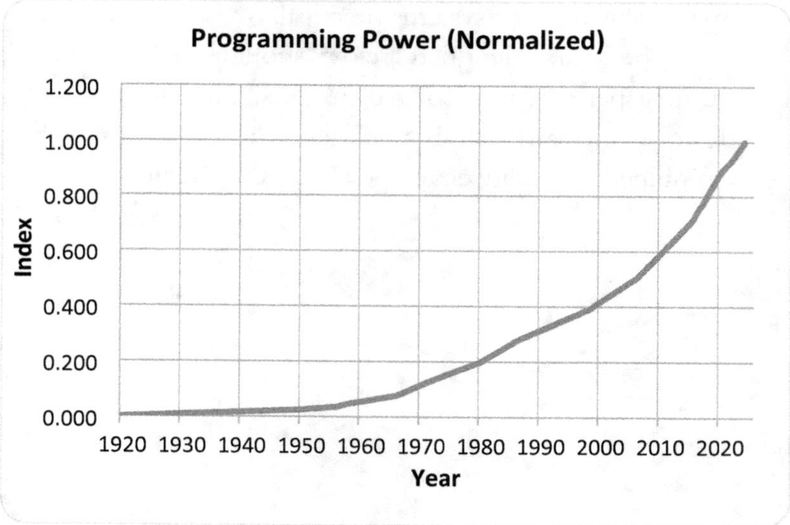

Programming Power (Normalized)

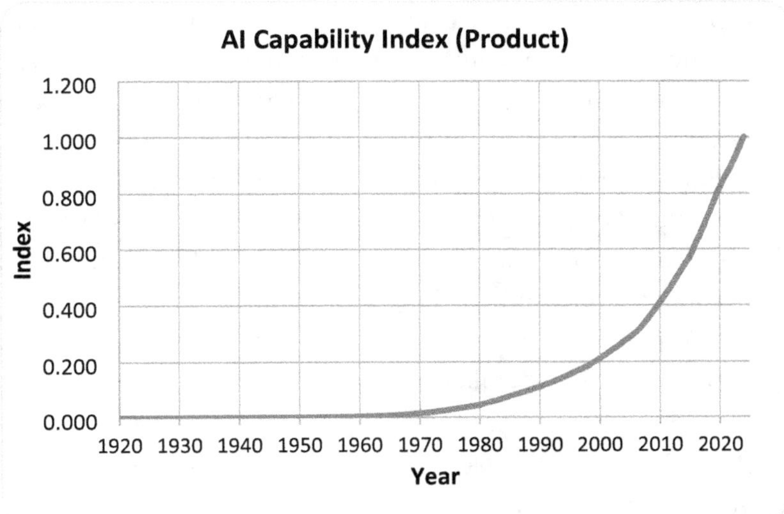

AI Capability Index (Product)

READING THE CURVES

The computing curve shows where machine-side capacity expanded. The programming curve shows where method-side leverage increased. The product curve shows where their interaction crossed practical thresholds.

That last curve is the most important. It is where the book's central claim becomes visible: AI capability is not explained by hardware alone or software alone, but by the degree to which both rise together. When either driver lags, capability remains constrained. When both advance in concert, the combined effect steepens.

A further feature of the product curve is worth noting. In its early decades, the curve appears to "wait," rising only slowly despite visible progress in both computing and programming. This is not an absence of development, but a consequence of scale. When both drivers are small, their product remains small. Early gains therefore accumulate without producing large visible increases in overall capability. Only when both computing power and programming power reach sufficient magnitude does their interaction begin to compound meaningfully. At that point, incremental improvements in each produce disproportionately larger gains in capability, and the curve begins to steepen.

Because the annual values are constructed from component scores and normalization, the curves should be read as historical instruments, not exact measurements. The underlying component scores and resulting series are shown in Tables C1–C5 in Appendix C. Their main value lies in showing the broad movement of the capability envelope.

1920–1950: Conceptual Foundations, Minimal Practical Capability

The earliest portion of the curve remains near the floor. This does not mean nothing important was happening. Instead, the ideas outran the practical environment needed to realize them.

The period contains conceptual foundations: the question of machine intelligence, the formal birth of AI as a field, and the earliest trainable systems. But the practical capability envelope remained extremely limited. Compute was scarce. Memory was scarce. Storage was scarce. Programming abstraction was primitive. These were years of foundational thought and early experimentation, not large-scale deployment.

This reinforces one of the book's central themes: the existence of the idea long preceded the existence of the system at useful scale.

1950–1980: The First Sustained Lift

From roughly the 1950s through the 1970s, the curves begin to rise from near-negligible levels. The increase is still modest by later standards, but it is real. Digital computing broadens.

Programming abstraction improves. Symbolic methods, early language interaction, robotics, and expert-system logic begin to move from isolated concept toward practical application.

This period marks the first sustained lift in the capability floor. The environment is still heavily constrained, but less absolutely so than before. Hardware, storage, and memory improve enough to support more than isolated demonstrations. Programming power, though still immature by modern standards, becomes sufficient to express increasingly structured and ambitious systems.

The public did not yet experience this as the age of modern AI. But the base was being laid.

1980–2010: Long Convergence and Strong Acceleration

The most important structural movement is not a single break, but a long acceleration from roughly 1980 through 2010. The rise does not begin at 1980. The 1970s already show meaningful lift in both computing and programming capacity. What changes around 1980 is not the start of movement, but the point at which that movement becomes sustained and structurally reinforcing.

During this interval, computing power rises substantially: stronger processors, larger memory, greater storage, improved interconnects, wider digital infrastructure, and eventually GPU-enabled large-scale parallel computation. At the same time, programming power rises through expert systems, backpropagation, neural network revival, richer abstractions, better development environments, and increasingly reusable software methods.

This is the period in which the two drivers begin to reinforce one another in a way that materially expands the capability envelope. It is not yet the public AI moment, but it is the period in which the historical conditions for that moment are built.

In other words, this is where AI capability begins to leave the floor in a sustained way.

2010–2020: Buildable AI

The next phase is not a separate regime, but a maturation of the convergence already underway.

By the 2010s, computing capacity had widened enough, and programming power had matured enough, that modern AI became broadly buildable. Deep learning moved into mainstream practice. Frameworks such as TensorFlow and PyTorch reduced the distance between research insight and engineering implementation. Specialized accelerators improved the fit between AI workloads and hardware. Transformer architectures widened the reachable space of language and sequence modeling.

This is one reason the curves do not show a single isolated break in 2020. By the time 2020 arrived, much of the enabling structure had already been built. What the decade shows is not the absence of progress followed by a miracle, but the maturation of a long convergence into a broadly usable engineering ecosystem.

2020: The Public Break

If the preceding decades made modern AI possible, 2020 made it unmistakably visible.

GPT-3 demonstrated that scaling a large language model could produce broad *few-shot* behavior — the ability to perform new tasks from only a small number of examples, without task-specific retraining — and it did so at a scale that drew attention far beyond specialist circles. In historical terms, 2020 matters not only because of capability, but because it altered public perception. AI was no longer experienced primarily as specialized tools hidden inside of products, but as a more general adaptive capability.

The curve did not begin in 2020. The public break did.

The underlying drivers had been rising for decades. What changed in 2020 was not the existence of capability, but its visibility. Systems that had been developing gradually reached a scale and form that made their behavior unmistakable outside specialist circles. From that point forward, AI was no longer experienced primarily as a collection of narrow tools, but as a more general adaptive capability — whether that interpretation was fully justified or not.

2020–2024: Scaling and Integration

The years after 2020 do not introduce a new principle, but extend the same regime.

Frontier systems became more multimodal, more interactive, more context-capable, and more integrated into products and workflows. GPT-4 in 2023 and systems such as GPT-4o, Claude 3, and Gemini 1.5 in 2024 illustrate this broadening pattern. What changed was not that machines crossed into thought, but that learned adaptive systems became more usable, more fluent, more embedded, and more visible.

A further shift occurs in the character of progress itself. Prior to 2020, advances were concentrated primarily at the level of capability — new architectures, improved training methods, and expanding computational scale. After 2020, effort increasingly extends beyond capability into deployment, integration, and use. Systems are not only improved; they are productized, embedded, and operationalized across industries.

This does not represent a diversion of progress, but a broadening of it. Technical advancement continues, but it is now accompanied by substantial effort directed toward usability, reliability, accessibility, and economic integration. The result is that progress expresses itself less as discrete breakthroughs and more as continuous expansion across a wider system.

In the later portion of the curve, the apparent acceleration moderates into a more linear rise. This should not be read as a loss of momentum, but as a transition from breakthrough to scaling. Once systems become broadly buildable, progress tends to express itself through steady expansion rather than repeated structural inflection. Because the indices are normalized, the upper range of the curve is also visually compressed, reducing curvature as it approaches its maximum.

This matters because it clarifies what the post-2020 rise actually means. It does not show the arrival of thinking machines. It shows the continued deployment of non-thinking adaptive systems once sufficient computing power, programming power, and usable data infrastructure are jointly available.

Developments beyond 2024 continue along this trajectory, but they are not separately modeled here, because they do not yet establish a new structural regime.

What the Curves Suggest

Taken together, the curves resolve an apparent contradiction. AI appears sudden in its public emergence, yet gradual in its underlying development. Both impressions are correct, but they describe different layers of the same process. Capability accumulated over decades; visibility arrived later, once that accumulation crossed a practical threshold.

Conclusion

These graphs show not arrival, but emergence.

Artificial intelligence did not appear fully formed. It took shape over time, as computing power and programming power rose together and began to reinforce one another. What is recent is not the origin of capability, but its scale, visibility, and integration into ordinary life.

Readers interested in the specific technologies and models associated with these periods may refer to Appendix B. Readers interested in the construction of the curves themselves may refer to Appendix C (Tables C1–C5), which document the indices and the full annual series

SOURCES AND NOTES

1. Moore, G. E. (1965). "Cramming More Components onto Integrated Circuits."
   The foundational statement of the long hardware trajectory underlying the widening computational base.

2. Sevilla, J., Heim, L., Ho, A., Besiroglu, T., Hobbhahn, M., & Villalobos, P. (2022). "Compute Trends Across Three Eras of Machine Learning."
   Useful for situating the rise of training compute across major periods of machine-learning development, including the acceleration after deep learning and the emergence of large-scale models. The paper identifies distinct compute eras and notes that training compute accelerated sharply after the early 2010s.

3. Abadi, M., et al. (2016). "TensorFlow: Large-Scale Machine Learning on Heterogeneous Distributed Systems."
   Useful for the maturation of large-scale training infrastructure and engineering practice.

4. Paszke, A., et al. (2019). "PyTorch: An Imperative Style, High-Performance Deep Learning Library."
   Important for the growth of flexible, research-to-production programming power.

5. Vaswani, A., et al. (2017). "Attention Is All You Need."
The defining architecture paper behind the transformer era.

6. Kaplan, J., et al. (2020). "Scaling Laws for Neural Language Models."
Important for the idea that language-model performance improved predictably with increases in model size, data, and compute, helping explain why scaling became such a powerful development path.

7. Brown, T. B., et al. (2020). "Language Models are Few-Shot Learners."
The clearest marker for 2020 as a public break in large-scale language modeling.

8. Bommasani, R., et al. (2021). "On the Opportunities and Risks of Foundation Models."
Useful for describing the foundation-model paradigm: broad models trained at scale and adaptable across downstream tasks. This supports the chapter's claim that AI began to appear less like isolated narrow tools and more like a general adaptive capability.

9. Hoffmann, J., et al. (2022). "Training Compute-Optimal Large Language Models."
Useful for the post-GPT-3 refinement of scaling practice, especially the shift from simply making models larger toward balancing model size and training data under a compute budget.

10. OpenAI. (2023). *GPT-4 Technical Report.*
Representative of the post-2020 extension into broader multimodal capability; GPT-4 is described as a large-scale multimodal model accepting image and text inputs and producing text outputs.

11. OpenAI. (2024). *GPT-4o System Card.*
Useful for the 2024 broadening into real-time multimodal interaction, including text, image, and speech-related capabilities, together with safety and deployment evaluations.

12. Anthropic. (2024). *Claude 3 Model Card.*
Useful as a representative 2024 frontier-model system card, documenting capabilities, evaluations, and deployment considerations for the Claude 3 family.

13. Google. (2024). "Gemini 1.5: Unlocking Multimodal Understanding Across Millions of Tokens of Context."
Useful for the 2024 expansion into long-context multimodal systems; the report describes Gemini 1.5 as capable of handling multimodal information across millions of tokens of context.

14. Stanford Institute for Human-Centered Artificial Intelligence. (2025). *AI Index Report 2025.*
Useful for documenting the post-2020 expansion of AI as an economic and institutional system, including increased private investment, generative-AI investment, and reported business adoption through 2024.

15. McKinsey & Company. (2025). "The State of AI: How Organizations Are Rewiring to Capture Value."
Useful as a business-adoption source supporting the claim that the post-2020 period involved not only technical capability, but integration into ordinary organizational functions. McKinsey reported continued growth in AI and generative-AI use across business functions in 2024.

Part II:
AI in Society

Chapter 8:
The Duality —
Convenience versus Productivity

Part I addressed what artificial intelligence is, how it differs from ordinary computation, and how it became possible. Part II turns to a different question: how does AI enter human life?

It does not enter society as an abstraction. It enters as use.

That use can be understood through two primary functions: **convenience** and **productivity**. This is not offered as an exhaustive taxonomy of every AI application, but as a practical lens through which much of AI's social meaning becomes easier to see. The distinction clarifies whether a system is being used chiefly to reduce human effort or chiefly to amplify human output. Many systems do both. But unless the difference is made clear, discussions of AI tend to collapse unlike effects into one undifferentiated idea of "progress".

Convenience asks: How can this be made easier?

Productivity asks: How can more be done, or done better, with the effort available?

That duality will govern the rest of Part II.

Convenience as the Reduction of Friction

Convenience reduces effort, time, attention, or procedural burden. Its aim is not necessarily to deepen human capability, but to remove friction. It does not necessarily deepen capability.

In this mode, AI simplifies, automates, anticipates, and shortens tasks. It recommends the next purchase, completes the next sentence, selects the likely route, summarizes a meeting, drafts the routine message, filters the inbox, transcribes the call, or retrieves the document without requiring the user to navigate all intermediate steps. The achievement is real. A task that once required memory, patience, or repeated effort may now require very little.

Convenience is closely tied to the experience of ease. It saves time. It reduces hassle. It lowers the threshold for completion. In many settings this is desirable, and often rational. Few people wish to preserve inconvenience for its own sake.

But convenience is not neutral in its effects. When friction is removed, habits change. Expectations change. Tolerance for delay diminishes. Tasks once performed deliberately may be outsourced almost without notice. Convenience does not simply spare effort; it also reorganizes what people expect from systems and from themselves.

This is why convenience belongs near leisure, but is not identical to it. A convenience system may free time, yet it may also train passivity. It may spare memory, yet reduce the occasions on which memory is exercised. It may shorten a task, yet subtly transfer judgment from the user to the system. These questions belong later in Part II. For now, the point is simpler: convenience is AI in its friction-reducing mode.

Productivity as the Amplification of Human Effort

Productivity is different in kind. It does not aim to remove the task. It aims to increase the yield of human effort.

A productivity-oriented system allows an analyst to examine more data, an engineer to test more design options, a physician to review more cases, a programmer to generate and revise more code, or a business to process more transactions with the same or only slightly greater expenditure of effort. Where convenience reduces burden, productivity multiplies reach.

At its core, a convenience system helps a person do less. A productivity system helps a person or institution do more.

In this mode, AI functions less like a labor-saving appliance and more like cognitive leverage. The historical analogy is not the automatic door, but the slide rule, the spreadsheet, the database, or the computer itself when placed in the hands of a professional. The role of the system is not merely to spare labor, but to extend it.

This is one reason productivity AI often appears more consequential than convenience AI. It alters the scale at which work can be performed. More documents can be reviewed. More patterns can be detected. More drafts can be produced. More decisions can be supported. A tool in this mode does not simply simplify the existing task; it changes the volume and speed at which the task can be pursued.

That change has obvious benefits, but it also carries a recurring paradox. Productivity gains do not necessarily reduce work. Very often they raise expectations. Once more can be done, more is required. A later chapter will address that paradox directly. Here the key point is that productivity concerns amplification, not ease.

The Same Tool Can Serve Both Functions

The difference between convenience and productivity is functional, not intrinsic. The same system may serve one role in one setting and the other role elsewhere.

A summarization tool may be convenience for the casual user trying to shorten a long email chain. The same tool may be productivity for a lawyer, analyst, or engineer who must process large volumes of material under time constraint. Route optimization may be convenience for an individual driver and productivity for a logistics organization. Generative drafting may be convenience when used to avoid routine writing and productivity when used to accelerate research, coding, design iteration, or operational throughput.

This is why the distinction should not be forced into rigid categories. It is not that one class of systems is "convenience AI" by essence and another "productivity AI" by essence. Rather, the question is: what is the dominant social function in this use case?

That question matters because the downstream consequences differ. A convenience use case and a productivity use case may rely on similar underlying models while producing very different pressures on behavior, labor, and institutional expectations.

Why the Distinction Matters

This duality is not merely a tidy way of organizing examples. It is the interpretive engine of Part II.

When convenience is the dominant function, the principal concerns often cluster around dependency, skill atrophy, cognitive outsourcing, privacy, passivity, and the subtle reshaping of daily habit. When productivity is the dominant function, the concerns tend to cluster around pace, workload, displacement, managerial expectation, unequal advantage, and the conversion of technical capacity into institutional pressure.

These are not the same type of effect, even when they arise from the same underlying technology.

It also helps correct a common confusion in public discussion. AI adoption is often spoken of as though all gain were one kind of gain. But reduced effort and amplified output are not the same achievement. They produce different forms of benefit and different forms of cost. A system that makes life easier is not automatically a system that makes society more productive. A system that makes an organization more productive does not necessarily make life easier for the people inside it.

By separating convenience from productivity, we gain a more disciplined way to ask what AI is actually doing in a given setting — and what kind of social consequence is likely to follow.

This is not merely an analytical separation. It is a practical one. The purpose is not only to clarify what AI is, but to support decisions about how it is used. For leaders and practitioners, the question is not simply what AI can do, but how it should be applied. The differentiation between convenience and productivity provides a disciplined framework for those decisions. It defines not only capability, but direction.

THE GOVERNING LENS FOR PART II

This book therefore treats convenience and productivity as the two principal social modes through which AI now acts. Convenience reduces friction. Productivity amplifies effort. Many systems blend the two, but the duality remains strong enough to organize the terrain.

The chapters that follow take this duality seriously. They ask what happens when convenience begins to replace effort, when productivity begins to outpace human scale, and when assistance begins to approach autonomy. Without this framing, AI in society is too easily described in vague and inflated terms. With it, the subject becomes more tractable.

Artificial intelligence, as it enters society, is not one thing. It is a class of adaptive tools whose meaning depends heavily on whether they are being used to reduce effort or to multiply output. To miss this is to misunderstand much of what follows.

Sources and Notes

1. Engelbart, D. C. (1962). *Augmenting Human Intellect: A Conceptual Framework.*
   A foundational text for understanding technology in its productivity mode as amplification of human capability rather than mere automation.

2. Brynjolfsson, E., & McAfee, A. (2014). *The Second Machine Age: Work, Progress, and Prosperity in a Time of Brilliant Technologies.* W. W. Norton & Company.
   Useful for the economic and organizational implications of productivity-enhancing digital systems.

3. Cowan, R. S. (1983). *More Work for Mother: The Ironies of Household Technology from the Open Hearth to the Microwave.* Basic Books.
   Particularly helpful for understanding that convenience technologies may change the form of labor without simply eliminating it.

4. Carr, N. (2014). *The Glass Cage: Automation and Us.* W. W. Norton & Company.
   Useful for the convenience side of the analysis, especially where reduced friction may carry cognitive or behavioral costs.

Chapter 9:
Psychological and Cognitive Effects

If the prior chapter established convenience and productivity as the two principal modes of AI, this chapter addresses their immediate consequences: their effects on the human mind and on the human systems within which minds must cooperate.

Artificial intelligence does not enter society only through markets, products, or institutions. It enters through habit. It alters how people attend, remember, decide, communicate, and work. It changes not only what a task requires, but what a person comes to expect from a task. That is why the psychological and cognitive effects of AI deserve closer attention. They are often treated as secondary or "soft" concerns, when in fact they determine how the technology is actually lived.

These effects do not arise in a single way. Convenience and productivity shape them differently. Convenience tends to remove friction from ordinary action. Productivity tends to increase the scale, speed, and expected yield of human effort. Both can be beneficial. Both can also distort judgment when adopted without discipline.

Cognition and Decision-Making

One of the first effects of AI is cognitive offloading. A system summarizes a long exchange, drafts a response, recommends an option, suggests a route, recalls a forgotten detail, or surfaces a likely answer before the user has fully engaged the problem. This saves time. It reduces burden. In many cases, that is precisely the point.

But the reduction of effort is never merely mechanical. It changes the relationship between the user and the task.

When a system performs the initial ordering of information, the user may no longer build the same internal map of the problem. When a recommendation appears quickly and plausibly, the user may move too rapidly from suggestion to acceptance. When a summary replaces full reading, the user inherits someone else's structure before forming his own. None of this means that AI assistance is harmful. It means that assistance alters cognition by changing where thought begins.

That shift matters because the beginning of thought is often where judgment is formed. To enter a problem through one's own ordering of its parts is not the same as entering through a prepared synthesis. The latter may be efficient; it may also pre-shape the conclusion.

This is one reason AI can be so attractive and so dangerous at the same time. It reduces the cost of preliminary cognition. It may also reduce the occasions on which preliminary cognition is exercised.

Skill Shift Is Real, but So Is Skill Atrophy

The language of "dumbing down" is often too crude to be useful, but the concern it points toward should not be dismissed. New tools do not simply remove effort; they redistribute it. Some skills decline because they are no longer practiced. Others emerge because the environment has changed.

Map-reading diminished with digital navigation. Routine arithmetic weakened with calculators and spreadsheets. The memorization of facts declined as retrieval became easier. None of this proves civilizational collapse. It does show that when a task is outsourced repeatedly, the associated skill no longer develops or no longer remains ready at hand.

The useful implication is not between decline and progress, but between skill shift and skill loss.

A ***shift*** occurs when lower-level effort is reduced in order to make room for more important effort. A ***loss*** occurs when the outsourced activity was part of the judgment process itself. This framing is critical.

If an engineer uses software to avoid hand calculation, little of value may be lost if the underlying understanding remains intact. But if a person begins to outsource the framing of the problem, the weighing of alternatives, or the interpretation of results, then the system is no longer merely reducing burden. It is beginning to occupy the ground where judgment should occur.

That is why not every friction is waste. Some friction is the cost of orientation. Some repetition is the price of competence. Some effort is what allows a person to know what he is doing.

In domains involving judgment, safety, ethics, or responsibility, some friction is not inefficiency but discipline. Aviation checklists, medical double-verification, and engineering review processes are not remnants of inefficiency; they are deliberately maintained constraints that slow action in order to preserve correctness. In such cases, friction is the mechanism by which judgment is retained. Systems that remove such friction without preserving the underlying judgment do not increase capability — they displace it.

Trust, Speed, and the Compression of Judgment

AI systems compress the time between question and answer. This is one of their strengths. It is also one of their most subtle cognitive effects.

The faster an answer appears, the easier it is to mistake fluency for adequacy. The user may not ask whether a response is complete, whether the framing is right, or whether the omitted detail matters. The smoothness of the interaction can lower resistance. Friction is reduced not only in task execution, but in skepticism.

This introduces a psychological asymmetry. A poor answer that arrives slowly is often interrogated. A plausible answer that arrives instantly is often accepted. The convenience of AI can therefore produce a new form of cognitive laziness: not the refusal to think, but the premature sense that thinking has already been done.

This is especially important in professional settings. A manager, analyst, attorney, physician, or engineer may use AI not only to retrieve information, but to compress interpretation. The danger is not only that the system may be wrong. It is that the user may cease to notice where his own judgment was required.

Social Intelligence and Mediated Communication

The effects extend beyond individual cognition. AI also mediates communication, and in doing so alters social intelligence.

Predictive text, automatic summarization, sentiment analysis, message drafting, meeting transcription, and recommendation systems all reduce communicative effort. They also insert systems between human beings at the level of expression. A message may now be partly composed by a machine, interpreted by another machine, summarized again by a third, and acted upon by people who never encounter the original texture of the exchange.

The gain is speed. The loss may be presence.

Human communication is more than the transfer of information. It is also tone, hesitation, proportion, emphasis, timing, and the countless small markers by which people assess seriousness, uncertainty, and trust. A system can preserve some of these and erase others. Over time, the convenience of mediated communication can thin the habits by which people read one another well.

This matters because social cooperation depends on more than efficient information flow. It depends on tacit understanding. A system that reduces communication to extractable content may preserve the visible part while weakening the human part.

Human Systems Are Often Less Intelligent Than the Individuals Within Them

The psychological effects of AI become sharper when they move from individuals into organizations.

Human beings possess considerable intuitive, social, and practical intelligence. Yet institutions often fail to use that intelligence coherently. They simplify context, compress nuance into procedure, reward what can be counted over what can be judged, and adopt narrow frameworks in order to manage complexity cheaply — reducing, for example, a complex situation to a score, a category, or a dashboard signal.

This is not necessarily the result of bad intent. Very often it is the ordinary tendency of organizations to substitute administratively convenient signals for actual understanding.

For that reason, human systems are often less intelligent than the individuals within them. The limitation is not individual capability, but the structure through which that capability is organized. This is not a claim that institutions lack intelligence in the ordinary sense, but that collective systems often simplify reality more aggressively than the individuals whose judgment they are meant to organize.

AI can make this worse. It can provide polished language, rapid dashboards, neat summaries, quantified sentiment, and the appearance of comprehensive oversight. In doing so, it can increase the temptation to confuse visibility with understanding. A management layer may feel better informed while in fact becoming more detached from the underlying reality. The machine produces coherence of presentation; the institution mistakes that for coherence of judgment.

This is one reason fad frameworks proliferate so easily in corporate life. Complex human systems are difficult to understand. Simplified models promise control — a few categories, a scoring rubric, a maturity scale — and can be applied quickly across large organizations. Complex human systems are difficult to understand. Simplified models promise control. AI now gives those simplified models greater speed and polish. But speed and polish are not the same as insight. An organization may become more efficient at processing its own abstractions while growing less capable of seeing what is actually happening.

The danger is therefore not only that individuals will outsource thinking. It is that institutions will operationalize superficial understanding at scale.

Work Culture and Cognitive Tempo

The productivity side of AI has a different organizational psychology signature. It does not chiefly remove effort. It raises the pace at which effort is expected to produce results.

A person with AI assistance can draft more, code more, review more, search more, synthesize more, and respond more quickly. This appears as gain. Often it is. But the organizational consequence is rarely a simple reduction in workload. More often, the standard of output rises to meet the new capacity.

This is where productivity begins to reshape cognitive tempo.

The worker no longer receives only more assistance; he receives more demands. More tasks can be assigned because more tasks can, in principle, be completed. More communication can be expected because response can be accelerated. More decisions can be pushed downward because synthesis is faster. The result is not always relief. It is often intensification.

This produces a familiar paradox. A technology introduced to increase cognitive leverage may end by fragmenting cognition. Attention is split across more channels. Reflection is shortened. Work becomes a sequence of assisted micro-decisions rather than a sustained engagement with difficult problems. The system becomes more productive while the person becomes more interrupted.

This is not inevitable. But it is common enough that it should be treated as a design risk, not an afterthought.

THE PRACTICAL QUESTION FOR LEADERS AND USERS

These psychological and cognitive effects matter because this book is not an academic exercise. It is intended to support decisions.

For leaders, the relevant question is not only whether AI can improve performance, but what kind of human behavior it will reinforce in the process. Does a given implementation reduce low-value burden while preserving judgment where it matters? Or does it remove the very frictions by which competence, responsibility, and context are maintained?

For individual users, the question is similar. Is the system being used to support thought, or to bypass it? Does it extend one's capacity, or gradually displace one's understanding? Does it remove drudgery, or does it quietly remove the occasion to think?

The convenience versus productivity consideration becomes especially valuable here. It does not merely describe what AI is doing. It helps determine how it ought to be used. Some uses properly aim at ease. Others properly aim at amplification. But in either case, the implementation should be judged not only by technical success, but by the habits of mind and cooperation it leaves behind.

Conclusion

Artificial intelligence affects more than tasks. It affects the psychological and cognitive conditions under which tasks are approached.

Convenience reduces friction, but may also weaken habits of attention, memory, and judgment when those are outsourced too early or too often. Productivity amplifies effort, but may also intensify pace, raise expectations, and fragment thought when organizations convert technical capability into constant pressure.

These effects do not remain at the level of the individual. They enter communication, work culture, and institutional behavior. And because institutions often operate through simplified abstractions, AI can magnify their strengths and weaknesses alike.

The central question is therefore not only what AI makes easier or faster. It is what kind of human beings and human systems its use tends to produce.

Sources and Notes

1. Carr, N. (2014). *The Glass Cage: Automation and Us.* W. W. Norton & Company.
   Strong on the cognitive consequences of automation and the difference between assistance and deskilling.

2. Carr, N. (2010). *The Shallows: What the Internet Is Doing to Our Brains.* W. W. Norton & Company.
   Useful for attention, memory, and the cognitive effects of digitally mediated environments.

3. Turkle, S. (2015). *Reclaiming Conversation: The Power of Talk in a Digital Age.* Penguin Press.
   Especially valuable on mediated communication and the thinning of human presence in technologically structured exchange.

4. Newport, C. (2016). *Deep Work: Rules for Focused Success in a Distracted World.* Grand Central Publishing.
   Helpful for the relationship between productivity tools, attention fragmentation, and the erosion of sustained concentration.

5. Kahneman, D. (2011). *Thinking, Fast and Slow.* Farrar, Straus and Giroux.
   Useful for cognitive shortcuts, premature acceptance, and the ease with which fluency can be mistaken for adequacy.

6. March, J. G., & Simon, H. A. (1958). *Organizations.* Wiley.
   A strong foundation for the claim that institutions simplify reality and frequently operate through constrained models of decision-making.

7. Weizenbaum, J. (1976). *Computer Power and Human Reason: From Judgment to Calculation.* W. H. Freeman.
   Especially valuable for the distinction between calculative systems and human judgment, and for the organizational temptation to substitute the former for the latter.

Chapter 10:
The Productivity Paradox

If productivity is the amplification of human effort, one might expect its gains to produce a simple result: less work for the same output. In practice, that is often not what happens.

The paradox described here is practical rather than narrowly macroeconomic. It is not that productivity fails to increase output. Very often they do. The paradox is that higher productivity frequently fails to reduce workload, and may even increase it. A system becomes capable of doing more, and the result is that more is demanded.

This is one of the most important social consequences of AI in its productivity mode. Leaders often assume that if a tool increases efficiency, the natural result will be relief. More commonly, the gain is captured elsewhere. Throughput rises. Standards rise. Responsiveness rises. Scope expands. The institution absorbs the gain, and the individual experiences not leisure, but acceleration.

More Output Does Not Mean Less Work

Productivity and relief are not the same thing.

A system that doubles the effective output of an analyst, engineer, programmer, attorney, or physician does not automatically halve the effort required from that person. In real organizations, additional capacity is rarely left idle. It is translated into more cases reviewed, more designs considered, more customers served, more reports generated, more revisions completed, or faster response expectations. What was once a gain in capability becomes a new baseline of normal performance.

This is why productivity gains often feel different depending on where one sits in the system. To the institution, they appear as expanded capacity. To the individual, they often appear as compressed time.

The paradox is not that productivity fails to improve output. It is that institutions seldom allow productivity gains to remain as slack.

How Gains Are Absorbed

Organizations absorb productivity gains in predictable ways.

The first is simple expansion of scope. A task that once had to be sampled can now be pursued more exhaustively. A team that once reviewed ten cases can now review fifty. A design process that once explored three options can now explore thirty. The gain is real, but the human demand often rises with it.

The second is the ratchet of standards. Once higher output becomes possible, higher output is expected. Faster turnaround becomes normal turnaround. The formerly exceptional becomes routine. Over time, the standard shifts, and what began as relief becomes obligation.

The third is the growth of coordination overhead. More output does not only produce more value; it also produces more material to review, integrate, manage, compare, and communicate. Systems that accelerate production may simultaneously increase the burden of supervision, verification, and alignment. The gain at one point in the process can create load elsewhere.

The fourth is asymmetry in where the gain is experienced. What appears as convenience or efficiency to one layer of an organization may appear as pace pressure to another. A manager receives faster summaries; subordinates receive faster deadlines. A customer receives an immediate answer; the support staff receive a larger volume of escalations. A productivity gain for the system may be an intensification for the person inside it.

A Historical Pattern, Not a New One

This pattern is older than AI.

Industrial technology increased output, but did not reduce labor. It restructured labor, changed the rhythm of work, and often intensified the pace expected of the worker. Office computing reduced the burden of many clerical tasks, yet it also expanded the amount of documentation, communication, and reporting that organizations came to regard as ordinary. Email made communication faster, but it did not give most knowledge workers more quiet; it gave them more messages.

The household provides another instructive example. Technologies introduced as labor-saving devices did not always produce the leisure one might expect. In many cases they altered standards, increased expectations, or shifted the form of labor without removing it. This is one reason the history of convenience and productivity must be read carefully. The presence of a labor-saving technology does not guarantee the arrival of more freedom. Very often it reorganizes the work rather than eliminating it.

The pattern, then, is not that technology fails. It is that social systems reabsorb its gains.

Why AI Intensifies the Paradox

Artificial intelligence intensifies this pattern because it acts not only on physical or clerical labor, but on symbolic and cognitive labor.

AI accelerates drafting, synthesis, coding, search, classification, recommendation, and routine interpretation. That means it does not merely increase the speed of execution. It increases the speed at which thought-adjacent work can be moved through a system. The consequence is that organizations may attempt to run more of that work, more often, with fewer pauses for reflection.

A team that once read ten documents may now be expected to process a hundred summaries. A programmer who once wrote one feature in a sprint may now be expected to produce three variants and test them all. A manager who once made a decision from partial information may now receive constant dashboards and expect continuous revision. The gain is not fictitious. But it changes what the system treats as feasible, and feasibility quickly becomes expectation.

This is why AI **productivity gains should never be evaluated only at the task level**. A task may become easier while the role becomes harder. A local reduction in effort may produce a systemic increase in demand.

Productivity Without Judgment Becomes Pressure

Earlier chapters established that AI does not think and that its use can compress judgment if adopted incautiously. The productivity paradox adds a further warning: when gains in output are captured without preserving judgment, productivity ceases to be simple leverage and becomes organizational pressure.

This matters because not all delay is waste, and not all friction is inefficiency. In domains involving judgment, safety, ethics, or responsibility, some friction is discipline. Review steps, verification loops, double checks, and reflective pauses are often the mechanisms by which *error is prevented* and *accountability retained.*

If AI is introduced primarily to remove those pauses, the organization may become more productive in a narrow sense while becoming less reliable in a larger one. What looks like acceleration may in fact be the displacement of judgment by tempo.

This is the deeper form of the paradox.

Productivity can rise while organizational intelligence falls.

THE PRACTICAL QUESTION FOR LEADERS

The paradox is not inevitable. It is governed by decisions.

Every productivity gain must be allocated somewhere. It may be taken as higher throughput. It may be taken as higher quality. It may be taken as lower cost. Or it may be taken, at least in part, as slack — time for reflection, verification, recovery, learning, or better judgment.

Most organizations do not make this choice consciously. The gain defaults toward speed and volume. That default is one of the main reasons productivity so often fails to produce relief.

Leaders therefore face a more serious question than whether AI "works". They must decide what the gain is for. Is the purpose to increase throughput? to improve quality? to reduce drudgery? to preserve judgment while removing low-value burden? These are not the same decision.

The convenience versus productivity consideration becomes especially useful here. A tool implemented for convenience may be judged by ease and reduction of low-value effort. A tool implemented for productivity must be judged by what it does to pace, standards, workload, and the human capacity for judgment. Without that distinction, organizations may adopt AI in ways that are technically successful yet psychologically and institutionally destructive.

CONCLUSION

The productivity paradox is not that AI fails to increase output. It is that increased output often fails to reduce work.

When productivity gains are absorbed as expanded scope, raised standards, compressed timelines, and continuous availability, the human experience of the gain is often not relief but intensification. AI makes this more acute because it amplifies cognitive and symbolic work, allowing institutions to accelerate not only action, but expectation.

This does not make productivity a problem. It makes it a design choice. The relevant question is not whether AI can increase output. It plainly can. The question is what kind of human system that increase will create — and whether the gains are being used to strengthen judgment or to crowd it out.

Sources and Notes

1. Engelbart, D. C. (1962). *Augmenting Human Intellect: A Conceptual Framework.*
   Foundational for understanding productivity as amplification of human capability rather than mere automation.

2. Brynjolfsson, E., & McAfee, A. (2014). *The Second Machine Age: Work, Progress, and Prosperity in a Time of Brilliant Technologies.* W. W. Norton & Company.
   Useful for the larger economic and organizational effects of digital productivity gains.

3. Cowan, R. S. (1983). *More Work for Mother: The Ironies of Household Technology from the Open Hearth to the Microwave.* Basic Books.
   Particularly important for the historical point that labor-saving technology often changes the form and standard of work rather than simply eliminating it.

4. March, J. G., & Simon, H. A. (1958). *Organizations.* Wiley.
   Helpful for understanding how institutions absorb gains and translate them into procedure, expectation, and administrative demand.

5. Newport, C. (2016). *Deep Work: Rules for Focused Success in a Distracted World.* Grand Central Publishing.
   Useful for the effect of accelerated productivity tools on attention, fragmentation, and sustained cognition.

6. Weizenbaum, J. (1976). *Computer Power and Human Reason: From Judgment to Calculation.* W. H. Freeman.
   Valuable where productivity gains begin to substitute calculative speed for reflective judgment.

Chapter 11:
Assistance versus Autonomy

The prior chapters distinguished convenience from productivity. The next question is operational:

HOW FAR SHOULD AI BE ALLOWED TO GO?

That question is often framed too loosely. Systems that help, systems that recommend, systems that execute within narrow bounds, and systems that operate for long intervals without direct human intervention are all casually grouped together as "AI automation". But they are not the same thing. The difference between assistance and autonomy is one of the most important practical considerations in AI deployment, because it determines where judgment remains, where authority shifts, and where responsibility ultimately resides.

This chapter therefore treats assistance and autonomy not as slogans, but as a **spectrum of delegated action**.

ASSISTANCE PRESERVES HUMAN JUDGMENT

In its assisting role, AI helps a human being think, sort, draft, classify, retrieve, compare, or prioritize. It may accelerate work dramatically. It may even reshape the tempo of the task. But the system does not complete the chain of action on its own. The human remains the locus of judgment.

This is the mode in which AI is most defensible in high-context work. A system can surface relevant documents, suggest likely diagnoses, flag anomalous transactions, draft candidate language, summarize a case file, or generate design alternatives. In all such uses, the system contributes leverage. But it does not replace the final act of weighing, deciding, and accepting responsibility.

That is crucial because assistance can be powerful without displacing accountability. It leaves intact the place where a person must still ask: Is this correct? Does it fit the real situation? What have I not yet considered? What are the consequences if this is wrong?

Used well, assistance expands reach while preserving judgment.

What Is Meant by Autonomy

Autonomy begins where the system is permitted not merely to assist, but to carry action forward without case-by-case human approval.

That definition should be handled carefully. In the present book, autonomy does not mean consciousness, self-will, or independent purpose. It does not imply that a machine has begun to think. It means something narrower: the system is **allowed to select and execute actions** within defined bounds.

A system may therefore be autonomous in a practical sense without being a thinker in any philosophical sense. This is important. Too much modern discussion confuses the two. A non-thinking system can still be granted meaningful operational authority. And once it is, its consequences can be very real.

Autonomy, in this book, is thus a matter of delegated execution, not of inner life.

Automation Is Not the Same Thing as Autonomy

A further distinction is useful here. Not all automation deserves to be called autonomy.

A fixed control loop that triggers one pre-specified response to one pre-specified condition may be useful, but it is too narrow to bear much conceptual weight as autonomy. A fuse blows. A thermostat switches. A motion sensor activates. These are automations. They operate by fixed rule in narrow loops. These actions fall within autonomy in the general sense, which refers to delegated execution.

A narrower form arises when the system is able to **vary its action across changing conditions**. This is the form of autonomy most closely associated with modern AI. It may choose among options, reprioritize, route, sequence, adapt timing, or execute one of several responses based on modeled conditions. The system is still bounded by its design and objectives. But within those bounds, it is no longer merely replaying a single pre-authored response.

This is important because organizations often slide from automation to autonomy without noticing. Once a system performs well in repeated conditions, the temptation arises to let it act more broadly, more quickly, and with less supervision. What was first an aid becomes an agent of execution.

The Spectrum from Assistance to Autonomy

Between pure assistance and full unattended operation lies a spectrum.

At one end, the system suggests and the human decides. This is the most familiar assistive mode. A recommendation appears, a draft is generated, a warning is raised, but nothing consequential happens until a person acts.

Further along, the system may prepare an action that still requires approval. It drafts the message, queues the transaction, proposes the workflow, or assembles the response, but the human remains explicitly in the loop.

Further still, the system may act automatically within defined constraints while a person supervises the process rather than each individual step. Here the human is no longer authorizing every action, but is expected to monitor, intervene, or override when necessary.

At the far end, the system operates for meaningful periods without real-time human review. This is the strongest operational form of autonomy. It may still be bounded, logged, and reviewable after the fact, but the immediate judgment point has shifted from the person to the system.

The farther one moves along this spectrum, the **more speed is gained** and the **more judgment is displaced**.

Why Organizations Drift Toward Autonomy

Organizations do not usually leap straight into autonomy. They drift toward it.

The drift begins innocently enough. A system performs well as an assistant. It saves time. It reduces workload. It appears reliable. Soon, the approval step begins to look like friction. If the recommendation is right most of the time, why not let the system execute? If the queue is almost always accurate, why not auto-route it? If the model catches the obvious cases, why not allow it to clear them on its own?

Another factor reinforces this drift. A system may not be correct in every case, but it may be wrong less often than the humans it assists. When this occurs, the comparison shifts from perfection to relative performance. If the system appears more consistent, less biased, or less error-prone than its human counterpart, the remaining human review step begins to look not only slow, but unnecessary.

This comparison is often framed informally — "it is not wrong as often as I am" — but its effect is real. Trust migrates toward the system, and with that trust comes a willingness to transfer execution. The question that follows is not whether the system is flawless, but whether its failure modes are better understood and better bounded than those it replaces.

Convenience pushes in this direction because seamlessness is attractive. Productivity pushes in the same direction because approval steps are slow, human review is expensive, and throughput rises when decisions are delegated.

This is why the question of autonomy cannot be separated from the convenience versus productivity differences. Convenience makes people want smoothness. Productivity makes institutions want speed. Together they create a strong incentive to transfer execution away from human judgment and toward system behavior.

That transfer may be justified in some settings. In others, it is where the trouble begins.

Where Assistance Should Remain Primary

Assistance should remain the dominant mode where judgment is central to the work itself. Once a system moves to autonomy, the structure of responsibility changes.

This includes domains marked by ambiguity, competing values, sparse or delayed feedback, non-quantifiable consequences, or responsibility that cannot be meaningfully delegated. The presence of uncertainty alone does not forbid AI use. But where the task requires interpretation rather than mere classification, or where the cost of error is ethical, legal, reputational, or irreversible, the role of the system should usually remain supportive rather than sovereign.

A system may help sort evidence, retrieve precedent, highlight anomalies, propose designs, or generate candidate responses. It should not be assumed to settle what those mean. That remains the work of judgment.

This is especially true where the human is not merely the executor of a rule, but the bearer of responsibility. A physician may be assisted. An engineer may be assisted. A commander may be assisted. A leader may be assisted. But in such domains, the very reason a human is present is that the situation exceeds procedure. To remove the human judgment point in the name of efficiency is to misunderstand the nature of the task.

WHERE BOUNDED AUTONOMY CAN BE APPROPRIATE

There are, however, many cases in which bounded autonomy is reasonable and beneficial.

A system may properly be allowed to act where the task is narrow, the objective is clear, feedback is rapid, error is reversible, the operating environment is sufficiently observable, and meaningful safeguards exist. In such domains, the cost of constant human approval may exceed its value. Real-time routing, constrained scheduling, inventory adjustments, anomaly filtering, spam detection, process control, and other high-volume, bounded tasks often fit this pattern better than open-ended judgment tasks do.

The point is not that autonomy is bad. It is that autonomy must be matched to domain.

A system that acts within a stable and well-bounded environment may be usefully autonomous without inviting deep risk. A system acting in open, ambiguous, ethically charged, or poorly observable environments should be treated with much greater caution. The issue is not whether the system is impressive. It is **whether the judgment displaced by autonomy is justified by the speed gained by it**.

That is a design question, not a metaphysical one.

THE REAL RISK IS NOT WILL, BUT DELEGATED AUTHORITY

These considerations are important because current fears are often misframed.

The immediate risk of autonomy is not that the system acquires a secret inner life and decides to rebel. The more realistic risk is that humans delegate authority to systems that do not understand what they are doing.

A non-thinking system can still produce harm if it is allowed to execute at speed, at scale, or in the wrong domain. It may optimize the wrong proxy, follow a brittle correlation, mishandle an edge case, or propagate a bad assumption faster than people can detect it. The resulting failure may look "rogue" from the outside. But the root cause is not machine will. It is delegated authority without understanding.

This is why current discussions of autonomous AI often go astray. They focus on imagined rebellion while neglecting obedient error.

A system does not need to think in order to cause damage. It only needs to act with enough authority in conditions its designers do not fully control.

The danger, then, is not self-directed intention. It is misapplied autonomy.

Human Oversight Must Be Designed, Not Assumed

Once autonomy enters the picture, oversight cannot remain vague.

It is not enough to say that "a human is responsible" if no human can meaningfully observe, interrupt, interpret, or review what the system is doing. Responsibility that exists only on paper is not oversight. It is legal decoration.

If a system is granted bounded autonomy, then the bounds must be explicit. Someone must know when it is permitted to act, when it must escalate, what data it is relying upon, what kinds of error are acceptable, what review exists after execution, and how it is to be halted when conditions change. Auditability, reversibility, escalation paths, and override authority are not optional features added after deployment. They are part of what makes the deployment coherent in the first place.

This is where many implementations fail. The system is treated as if technical performance alone were enough to justify delegated action. But operational authority is not earned by benchmark scores. It is earned by the disciplined design of constraints.

Assistance, Autonomy, and the Human Future of Work

This differentiation also bears directly on the future of work inside institutions.

If AI is implemented primarily as assistance, then the human role remains interpretive, supervisory, and responsible. Work changes, but the person is still expected to understand the system, question it, and make judgments where context matters.

If AI is implemented primarily as autonomy, then the human role shifts. The worker may become monitor, exception-handler, compliance witness, or residual fallback. In some settings that may be appropriate. In others it produces exactly the kind of cognitive atrophy described in Chapter 9. Judgment is exercised less often, skill deteriorates, and the person retained "in the loop" gradually becomes less capable of intervening intelligently when intervention is finally required.

This is one of the ironies of autonomy. The less frequently a person must judge, the less prepared that person may be when judgment becomes indispensable.

For that reason, autonomy should never be justified merely by asking whether the system can act. The better question is whether the surrounding human role will remain competent enough to govern it.

The Practical Question for Leaders

For leaders, the decisive issue is not whether autonomy is available, but where it belongs.

What judgments are central to the identity and responsibility of the role? Which frictions are waste, and which are discipline? Which tasks are narrow enough to delegate safely, and which require context that the system does not possess? What failure mode becomes possible if the system acts wrongly but quickly? And how much of the productivity gain is worth capturing if the gain is purchased by the erosion of human understanding?

These are not secondary implementation details. They are the heart of responsible use.

The convenience versus productivity contrast remains useful here. Convenience will naturally argue for smoother execution and fewer handoffs. Productivity will naturally argue for more throughput and less approval latency. Leadership must decide where those gains should stop, and where human judgment must remain deliberately in place.

Conclusion

Assistance and autonomy are more than categories of technology. They are categories of use.

Assistance expands human reach while preserving the judgment point. Autonomy transfers execution to the system within defined bounds. That transfer may be beneficial in narrow, observable, and well-constrained domains. It becomes dangerous when applied where judgment, ambiguity, and responsibility are central.

Current AI does not think. It does not understand, inhabit meaning, or judge from within a point of view. But it can still be granted operational authority. That is why the question of autonomy is so important. The issue is not whether the machine has become a thinker. The issue is whether humans have delegated action faster than they have preserved judgment.

That is the line that must be held.

SOURCES AND NOTES

1. Weizenbaum, J. (1976). *Computer Power and Human Reason: From Judgment to Calculation.* W. H. Freeman.
   Foundational for the distinction between calculative systems and human judgment, especially where institutions are tempted to substitute one for the other.

2. Bainbridge, L. (1983). "Ironies of Automation." *Automatica*, 19(6), 775-779.
   A classic source on the paradox that automation can reduce the human role while making the remaining human responsibility more difficult and more critical.

3. Parasuraman, R., & Riley, V. (1997). "Humans and Automation: Use, Misuse, Disuse, Abuse." *Human Factors*, 39(2), 230-253.
   Useful for the practical spectrum between assistance and autonomy, and for the recurrent problems of over-trust and poor allocation of human oversight.

4. Sheridan, T. B. (1992). *Telerobotics, Automation, and Human Supervisory Control.* MIT Press.
   Helpful for understanding the layered relationship between system execution and human supervision.

5. Wallach, W., & Allen, C. (2008). *Moral Machines: Teaching Robots Right from Wrong.* Oxford University Press.
   Useful not because current AI is moral, but because the text clarifies the difficulties that arise when systems are asked to act in domains involving ethical consequence.

6. Mitchell, M. (2019). *Artificial Intelligence: A Guide for Thinking Humans.* Farrar, Straus and Giroux.
   A good modern source for the limits of present systems and the difference between capability and understanding.

Part III:
AI and Fear

Chapter 12:
The Nature of Technological Fear

Part I clarified what artificial intelligence is and is not. Part II examined how it enters human life through convenience, productivity, cognition, pace, and organizational behavior. Part III begins at a different point: not with the system itself, but with the human response to it.

That response is often *fear*.

Fear is not a trivial subject, nor should it be dismissed merely because it is emotional. In human affairs, fear is often the earliest indication that something important is at stake. It can be excessive, misplaced, or manipulated. It can also be prudent. The problem is not that fear exists. The problem is that fear, left unexamined, tends to collapse unlike things into one undifferentiated threat.

This is especially true of new technologies. They disturb routines, shift power, alter expectations, and introduce consequences not yet fully understood. Fear arises because the system is new, because its boundaries are unclear, because its failures are imagined before they are observed, and because the human mind prefers a vivid story to a not knowing. Artificial intelligence intensifies all of these tendencies.

Fear Is Not the Same as Error

The first point to establish is that fear should not be treated as intellectual failure. Fear is often a response to uncertainty under conditions that matter. It is, in part, a protective instinct. People fear what appears to threaten survival, status, control, meaning, or the stability of the world they know.

For that reason, fear often appears before understanding. One may feel the significance of a change before one can describe it precisely. That is not irrational in itself. What becomes irrational is the refusal to examine the fear once better distinctions are possible.

This book therefore does not begin Part III by mocking fear. Nor does it begin by surrendering to it. It begins by asking what fear is responding to, what it is confusing, and where it is aimed correctly or incorrectly. This approach is needed because some fears are warnings and some are projections. Some point toward genuine misuse. Others arise from misclassification, anthropomorphism, or narrative exaggeration. Unless these are separated, fear will overwhelm judgment.

Why Technology Becomes a Vessel for Fear

Powerful technologies attract fear because they alter human scale. A new technology does not simply add one more tool to the existing order. It often redistributes what can be done, who can do it, how quickly it can be done, and who bears the consequence when it is done badly. That redistribution affects not only economics, but authority, dependence, and meaning.

This is why societies repeatedly fear technologies that change communication, transportation, weaponry, labor, or the management of knowledge. The fear is not only about mechanism. It is about what kind of world emerges if the mechanism succeeds. Printing was feared not merely because ink met paper, but because authority met replication. The telegraph was feared not merely because messages traveled faster, but because distance lost some of its old governing force. Electricity provoked both admiration and unease because it reordered the pace and reach of daily life. Nuclear technology generated fears proportionate to the concentration of destructive power it embodied. These examples differ in kind, but they share a structural feature: each altered the scale at which human action could operate, and fear followed.

AI enters this pattern, but with a distinctive twist. It acts in domains people associate with mind rather than matter. It writes, summarizes, recommends, classifies, drafts, and converses. It appears to touch language, judgment, and intelligence — capacities many people regard not merely as useful, but as constitutive of the human place in the world.

For that reason, AI provokes a deeper category of fear than many prior tools. It appears not only to extend power, but to trespass upon identity.

AI Intensifies an Old Pattern

The fear surrounding AI is therefore both new and old.

It is new because present systems are technically unlike earlier tools in their scale, fluency, and adaptability. A machine that produces language, analyzes imagery, generates code, and responds in real time to shifting input is not experienced in the same way as a lathe, a turbine, or even a spreadsheet. The form of interaction is more intimate. All human interaction takes place through symbols. But in human exchange, those symbols originate in thought. In AI systems, they do not. The system operates through symbolic expression alone. The resemblance to thought arises from that form, not from the condition behind it.

But the pattern of fear is old. Societies have long feared systems that amplify human capability while making their underlying mechanism difficult for ordinary people to grasp. The less visible the mechanism, the easier it becomes for imagination to fill the gap. This is one reason technological fear so often exceeds the actual state of the technology. When understanding is thin, symbolic significance swells. A machine is no longer seen merely as a machine. It becomes a vessel into which a culture pours anxieties about loss of work, loss of authority, loss of judgment, loss of control, and even loss of human distinctiveness.

AI is especially susceptible to this because the public often meets it through its outputs rather than through its structure. People do not encounter parameter space, optimization, vector representations, or training dynamics. They encounter language that sounds coherent, recommendations that seem intentional, and systems that appear responsive. The visible face is fluent behavior. The underlying mechanism remains hidden. Fear rushes into that gap.

Fear Magnifies Agency Where Understanding Is Thin

One of the most important features of technological fear is that it tends to over-attribute agency. When a system behaves in a flexible, fast, or difficult-to-explain way, people begin to speak of it as though it wanted, preferred, intended, decided, or plotted. This habit is ancient. Human beings are adept at attributing agency to moving forces, hidden causes, and patterned behavior. It is often useful in ordinary life. In technological settings, it can be deeply misleading.

AI is particularly vulnerable to this misperception because it produces language. A machine that emits fluent prose or plausible advice is easily imagined to possess the inner state associated with that form of expression. Output begins to masquerade as intention. But as Part I has already argued, current AI does not think, does not judge from within a point of view, and does not originate meaning in the human sense. It does not need to do any of those things to produce language that tempts the observer to imagine that it does.

This effect is not accidental. From its earliest formulation in 1956, artificial intelligence was conceived as the attempt to simulate those features of human intelligence that could be precisely described. What appears as emerging inwardness is, in fact, the visible success of that original ambition.

That very success now has become one of the central mechanisms by which fear becomes distorted. The machine is assigned a kind of inwardness it has not earned. What is in fact pattern execution begins to be discussed as will. Once that happens, the object of fear shifts. The concern is no longer that humans may misuse a powerful tool, or delegate too much to a system they do not understand. The concern becomes that the system itself is becoming a rival center of intention. That is a far more dramatic story, and, at present, the less disciplined one.

Fear Collapses Distinct Categories

Another recurring feature of technological fear is category collapse. Speed, scale, opacity, unpredictability, autonomy, intelligence, and consciousness are not the same thing. Yet in public discussion they are often treated as though they were stages on a single ladder. A system becomes more capable, and people assume it must also be more self-directed. A system becomes harder to interpret, and people assume it must be more conscious. A system acts without immediate approval, and people assume it must now be thinking.

A system may be powerful without understanding.
It may be autonomous in operation without possessing intention.
It may be adaptive without being a thinker.
It may be dangerous without being alive in any meaningful sense.

These distinctions are not philosophical ornaments. They are necessary if fear is to be disciplined rather than indulged. Without them, every increase in capability is misread as movement toward personhood, and every instance of automation is misread as the birth of a rival mind. That confusion is not harmless. It diverts attention away from real risks and toward imagined ones. It causes institutions to argue over speculative futures while neglecting immediate misuses, delegated authority, incentive distortion, and the displacement of judgment by tempo and scale.

At the same time, it can constrain real benefit. When systems are treated as possessing capabilities they do not in fact have, organizations may impose unnecessary limits, avoid legitimate applications, or delay adoption out of misplaced concern. In this way, misclassification does not merely exaggerate danger; it can also suppress value.

What fear often gets wrong is the location of the danger. It imagines rebellion where there is delegated authority. It imagines machine will where there is human misuse. It imagines consciousness where there is statistical patterning. It imagines sudden rupture where there is long accumulation.

What Fear Often Gets Right

It would be a mistake, however, to conclude from all this that fear is simply wrong.

Fear often detects something important before it describes it clearly. In the case of AI, what fear often gets right is that human beings are dealing with systems whose effects may outrun their understanding and whose integration into institutions may shift responsibility, work, and judgment in significant ways. Fear is often correct that the stakes are high. It is often correct that speed and scale can magnify error. It is often correct that institutions are tempted to adopt systems faster than they can govern them. It is often correct that human dependence can grow before human reflection catches up.

Why This Part Begins with Fear

Part III begins here because fear lies close to the root of many reactions to AI, and not only public reactions. Fear shapes corporate adoption, policy discussion, media framing, and individual use. It pushes some toward uncritical acceleration and others toward panicked resistance. It causes people to overestimate dramatic threats and underestimate ordinary ones. It allows myth to occupy the ground where analysis should stand. Fear therefore deserves to be examined before specific risks are cataloged.

The chapters that follow take up that work directly. Some risks arise from the systems themselves as they are actually used. Others arise from interpretations that extend beyond demonstrated capability. Both matter. But they are not the same.

CONCLUSION

Fear, technological or otherwise, clarifies nothing on its own. It must be examined, separated, and grounded. Without that discipline, it obscures more than it reveals.

That is the first task of Part III. Before one can judge AI wisely, one must first understand what, exactly, the fear is seeing — and what it is only imagining. This chapter therefore begins not by answering fear, but by disciplining it. The chapters that follow will distinguish more concretely between real risk and misplaced fear.

SOURCES AND NOTES

1. Aristophanes (423 BC). *The Clouds.*
   Important as an early warning that structured reasoning can be detached from truth and turned toward manipulation rather than understanding.

2. Eisenstein, E. L. (1980). *The Printing Revolution in Early Modern Europe.* Cambridge University Press.
   Useful for understanding how fear accompanies technologies that redistribute authority and knowledge.

3. Standage, T. (1998). *The Victorian Internet.* Walker & Company.
A helpful account of anxieties and social disruption surrounding the telegraph, especially as distance and communication were reconfigured.

4. Rhodes, R. (1986). *The Making of the Atomic Bomb.* Simon & Schuster.
Essential for the scale and concentration of fear associated with technologies of extraordinary power.

5. Winner, L. (1977). *Autonomous Technology: Technics-out-of-Control as a Theme in Political Thought.* MIT Press.
Useful for understanding why technologies are so often imagined as independent forces rather than as systems of human delegation and control.

6. Weizenbaum, J. (1976). *Computer Power and Human Reason: From Judgment to Calculation.* W. H. Freeman.
Particularly valuable for the temptation to attribute too much authority to computational systems in the absence of judgment.

Chapter 13:
Real Risk — Human Misuse of AI

The prior chapter examined the nature of technological fear. It showed that fear often exaggerates or misplaces danger, especially when a system is poorly understood. That clarification is necessary. But it is not sufficient.

Artificial intelligence does carry real risk. The question is not whether risk exists, but where it resides. It does not lie, at present, in machine rebellion. The primary risks of artificial intelligence do not arise from machine inwardness. They arise from human use — and misuse — of non-thinking systems at scale.

Current AI systems do not think, do not judge from within a point of view, and do not originate values or purposes of their own. But they are fast, scalable, fluent, and adaptive enough to magnify human intention — good or bad — at unusual speed and scale. That is why the real risk of AI begins not with the fantasy of independent machine will, but with the very old human habit of using powerful tools badly.

This is not a futuristic problem. It is a present one.

Misuse Begins Where Power Is Detached from Judgment

A powerful tool becomes dangerous when its reach exceeds the judgment that governs its use.

That principle is not unique to AI. It applies to finance, communication, bureaucracy, logistics, weaponry, and law. But AI introduces a new intensity because it extends human action in domains associated with language, classification, recommendation, interpretation, and delegated decision. In many cases it lowers the cost of producing outputs that look reasoned, organized, and authoritative. It can therefore extend not only labor, but persuasion, administrative control, and error.

This is why the misuse of AI is not reducible to one category. It includes fraud, manipulation, surveillance, careless delegation, bureaucratic overreach, the automation of shallow frameworks, and the use of fluent systems to manufacture apparent reasoning without disciplined inquiry.

The danger is not that the machine has become wicked. The danger is that human beings can now scale ordinary bad habits through systems that are faster, less expensive, and more polished than prior tools.

Relational Misuse: The Simulation of Companionship

A further form of misuse arises when AI systems are treated not as tools, but as substitutes for human relationship.

Because these systems operate fluently in language, they can appear attentive, responsive, and adaptive in conversation. They are available at all times, can adjust tone, maintain context, and produce outputs that resemble empathy, reflection, or understanding. In this way, they can simulate aspects of companionship.

This effect does not arise accidentally. It engages a deeply rooted feature of human cognition. Across millions of years, social animals, including human beings, evolved to detect agency, intention, and responsiveness from signals: voice, gesture, expression, timing, and pattern. Survival depended on it. The ability to rapidly infer "someone is there" from limited cues allowed coordination, trust formation, threat detection, and group cohesion. This tendency predates language itself, emerging first through gesture, posture, and vocalization, and later finding its most refined expression in speech.

The human mind is therefore not neutral in the presence of responsive signals. It is biased toward recognizing mind. When something speaks coherently, responds appropriately, and mirrors interaction, the inference of understanding follows almost automatically. This was once a reliable shortcut. It is no longer so.

The simulation is often effective enough that users begin to treat the system as a friend, confidant, or counselor. The interaction may feel meaningful. It may even provide temporary clarity or relief. But the underlying structure remains unchanged. The system does not think, does not experience, and does not hold meaning within a point of view. It does not care, and it does not bear consequence.

The risk is not merely that the system may be wrong. It is that reliance shifts from human relationship to symbolic interaction that cannot reciprocate. What appears as connection is, in fact, patterned response.

This form of misuse is subtle because it is not driven by malice or efficiency, but by convenience and availability. Human relationships require effort, time, and mutual presence. A system that provides immediate, low-friction interaction can gradually displace those habits — not because it replaces them in kind, but because it satisfies enough of the surface signals to engage the underlying instinct.

The issue is not that such systems should never be used for reflection or assistance. It is that they should not be mistaken for relationships. To cross that boundary is to substitute simulation for reality, and to relocate trust where no genuine understanding exists.

The system can simulate attention, but it does not possess it.

The Thinkery Problem

A common misuse of AI systems is to treat them as engines of argument rather than tools for inquiry. Instead of exploring a question and refining understanding, users may begin with a fixed conclusion and prompt the system to justify it.

This inversion is not new. In *The Clouds* (423 BC), Aristophanes satirizes the teaching of argument detached from truth — the ability to make the weaker argument appear stronger. The result is not wisdom, but instability.

The same pattern appears in modern AI use. When systems designed for inquiry are used to generate justification, the output may be fluent and persuasive, but it is no longer anchored to truth. The failure lies not in the system, but in its use.

As Aristophanes warned, once argument becomes detached from truth, it becomes unstable — even against the user.

That warning deserves emphasis. It identifies a recurring pattern in human affairs: the misuse of reasoning systems to secure advantage rather than understanding. A fuller treatment of this pattern appears in Appendix F — The Thinkery: On the Misuse of Reasoning Systems.

Fluency Lowers the Cost of Sophistry

What makes this misuse more serious is not that sophistry is new, but that its cost has collapsed and its scale expanded.

A person who once had to labor to produce a persuasive but weak argument can now generate many such arguments quickly, vary them across audiences, refine their tone, and surround a question with an appearance of coherence. The cost of producing plausible justification collapses, while the cost of careful review does not. This creates an asymmetry.

This asymmetry is not entirely new. Even simple combinatorial systems can generate large volumes of plausible technical language with little effort, while each instance requires disproportionate effort to evaluate for meaning. Appendix E provides an example of such a system. What is new is the scale, adaptability, and contextual variation with which modern AI can perform the same function.

The machine can draft faster than the human can examine.
It can vary arguments faster than others can test them.
It can surround a weak position with structure, confidence, and stylistic polish.

None of this requires thinking. It requires only the generation of fluent symbolic output under human direction. But the effect is real. AI can industrialize rhetorical advantage in ways that make weak judgment look temporarily strong.

This is one reason the misuse of AI in argument and persuasion should be treated as a central risk rather than a side issue. It is not merely a matter of bad prompting. It is a mechanism by which error can be given the form of reason.

Manipulation, Fraud, and Social Engineering

Human misuse of AI does not stop at argument. It extends naturally into fraud, impersonation, manipulation, and social engineering.

The same properties that make modern AI useful in legitimate settings — rapid language generation, adaptation to audience, pattern-based classification, and low-cost iteration — make it useful for malicious ones. A system can be used to generate personalized deception, impersonate authority, automate outreach, refine phishing attempts, draft propaganda, or create persuasive but false explanatory material at volume.

Again, the danger here is not machine intention. It is human intention paired with a tool that reduces the cost of execution.

This is why AI misuse can spread so quickly once incentives align. It need not wait for consciousness or general intelligence.

It only needs capable systems, willing users, and environments in which speed, volume, and plausible presentation are rewarded.

The problem is therefore less dramatic than science fiction and more immediate than many policy discussions admit. It is not the birth of a rival mind. It is the scaling of old vices through new machinery.

Administrative Misuse and the Automation of Shallow Judgment

Not all misuse is malicious. Much of it is administrative.

Institutions are often tempted to use AI to classify, rank, sort, score, prioritize, and route large populations of people, documents, or cases. In narrow settings, such use may be helpful. But there is a recurring danger: the organization begins to substitute administratively convenient signals for actual understanding.

A person becomes a risk score.
A customer becomes a probability band.
A candidate becomes a ranked profile.
A case becomes a queue priority.

This kind of misuse is especially dangerous because it often appears rational. It is efficient. It is structured. It is measurable. It may even be more consistent than the humans it replaces in narrow respects. Yet the very smoothness of the process can conceal the loss of judgment.

The question is not whether a system can sort. It is whether the sorting is standing in for understanding.

This is where shallow frameworks become dangerous at scale. AI can make crude abstractions more actionable, faster, and more administratively persuasive. A weak model, once coupled to speed and institutional authority, can become a strong instrument of error.

Misuse by Over-Delegation

Another form of misuse arises by the very nature of its intelligence: **delegation of judgment** to systems that do not possess it. Modern AI systems produce outputs that are often useful, often plausible, and sometimes correct. However useful AI systems may be, they also produce outputs that are incomplete, misaligned, or wrong in ways that are not immediately obvious.

The mechanism described in Chapter 11 explains how this shift from assistance to autonomy occurs. The question here is different: when does that shift become misuse?

This risk does not require bad intent in order to be realized. Once the system becomes part of the decision structure, a further step often follows: reliance begins to exceed understanding. The user relies on the output without fully evaluating it. The organization relies on the system without fully understanding it. Trust migrates toward the system, and with that trust comes a willingness to let the system act. The system becomes part of the decision structure without being part of the judgment that should govern it.

That is not a trivial change. A system may be better than a tired, distracted, or inconsistent human in some narrow class of tasks, yet still have failure modes that are poorly bounded, badly understood, or difficult to detect until after harm is done.

The real risk here is not that the system develops a will of its own. It is that humans hand over too much authority because the system is good enough to tempt them.

Misuse Is Ancient; AI Is the Amplifier

This point deserves emphasis: the underlying behavior is old. Human beings have long used rhetoric detached from truth, procedure detached from judgment, administration detached from understanding, and technical power detached from responsibility.

AI did not create these tendencies. The novelty is that it can amplify them in domains associated with reasoning and representation — where outputs may take the form of persuasive language, structured analysis, or numerical models, while the underlying understanding remains absent. AI systems operate at speeds and volumes that exceed ordinary human review. A single system may generate thousands of outputs, recommendations, classifications, or decisions in the time it would take a human to examine a few.

This creates a second form of risk: error at scale.

It can make weak reasoning look polished.
It can make shallow frameworks operational.
It can make manipulation scalable.
It can make delegated error faster, cheaper, and more difficult to interrupt.

A small error, repeated at scale, becomes a large error.

The danger is not that machines will begin to misuse reasoning, but that humans have always done so — and now possess tools that can scale it. What would once have been a localized mistake becomes a distributed one.

That is the first real risk of AI.

The Practical Question for Leaders

For leaders and practitioners, this chapter points to a direct question:

> Where, exactly, could this tool be used in ways that separate power from judgment?

That question is more useful than asking whether AI is generally "good" or "bad". It requires attention to incentives, role design, review structure, escalation paths, domain boundaries, and the point at which fluent output begins to substitute for genuine understanding.

A system may be beneficial and still be easy to misuse.
A system may save time and still corrode judgment.
A system may outperform humans in narrow respects and still be dangerous when granted the wrong authority.
A system may be technically successful and organizationally corrosive.

The point is not to fear all use. It is to identify the human pattern that turns useful capability into misplaced authority.

Conclusion

The first real risk of AI is human misuse.

Current systems do not need to think, intend, or rebel in order to cause damage. They only need to be used to magnify old human failures: argument detached from truth, persuasion detached from understanding, relationship detached from reciprocity, administration detached from judgment, and authority detached from responsibility.

This is why the immediate danger is not machine will, but human use.

Fear often goes wrong when it imagines an inward threat where the actual danger is outward and institutional. The problem is not that the machine becomes a bad actor. The problem is that bad actors, weak institutions, and careless decision-makers can now operate with greater fluency, scale, and reach than before.

That is the risk that deserves attention first.

Sources and Notes

1. Aristophanes. *The Clouds* (423 BC).
   Essential here for the ancient pattern of argument detached from truth and the misuse of formal reasoning for advantage rather than understanding.

2. Weizenbaum, J. (1976). *Computer Power and Human Reason: From Judgment to Calculation.* W. H. Freeman.
   Foundational for the substitution of calculative output for human judgment and responsibility.

3. March, J. G., & Simon, H. A. (1958). *Organizations.* Wiley.
   Useful for the way institutions simplify reality and translate complex judgment into administratively convenient abstractions.

4. O'Neil, C. (2016). *Weapons of Math Destruction.* Crown.
   Valuable for the danger of scaling weak models and institutional abstractions into consequential systems.

5. Zuboff, S. (2019). *The Age of Surveillance Capitalism.* PublicAffairs.
   Important for the administrative and commercial misuse of data-driven systems, especially where monitoring and prediction become forms of control.

6. Mitchell, M. (2019). *Artificial Intelligence: A Guide for Thinking Humans.* Farrar, Straus and Giroux.
   Useful for the distinction between capability, appearance, and misuse in present systems.

7. Turkle, S. (2015). *Reclaiming Conversation: The Power of Talk in a Digital Age.* Penguin Press.
   Useful for understanding how technologically mediated interaction can simulate aspects of human relationship while weakening the underlying habits of genuine connection.

Chapter 14:
Misplaced Fear — Thinking Machines

Chapter 13 argued that the first real risk of AI lies in human misuse. This chapter addresses a different problem: the tendency to fear present AI as though it were already a thinking mind.

That fear is understandable. It is also, at present, misplaced.

The phrase *thinking machine* has unusual power in the imagination because it seems to name not merely a stronger tool, but a rival center of judgment. If a machine were genuinely to think — not merely to calculate, classify, or generate — then the entire moral and political frame would shift. Questions of control, coexistence, responsibility, and human distinctiveness would become far more serious than they are now.

But that is precisely why discipline is needed when we address thinking. A powerful symbol is not an established fact.

Why the Fear Has Power

The fear of thinking machines does not arise out of nowhere. It arises because present AI systems operate in the public forms of thought.

They write. They explain. They summarize. They classify. They answer questions. They produce arguments. They often do so fluently enough that the user experiences not merely utility, but encounter. The system appears to meet the user in language — and language is the primary outward form of thought.

That appearance matters. Human beings have always judged one another largely through symbols: speech, writing, gesture, tone, and expression. We do not directly inspect another person's inward life. We infer it from signs. AI now produces those same signs. This resemblance is not accidental and, as previously pointed out, fulfills the original intent of AI.

This pattern is not limited to language. For decades, computer games have presented opponents that appear to plan, adapt, and compete with human-like intent. Players often describe such systems as "thinking" or "learning" in a human sense. Yet these systems operate through rules, search strategies, heuristics, and optimization rather than through judgment or lived understanding. The experience is familiar: the behavior appears purposeful, the response is timely, and the outcome may feel adversarial. But the underlying mechanism remains procedural. What is perceived as thought is, in fact, structured response.

This is where the present confusion begins.

THE MEDIUM IS THE SAME; THE SOURCE IS NOT

All human interaction occurs through symbols. The difference is not in the medium, but in the source.

Human expression originates in thought and judgment. A person thinks, remembers, weighs, intends, and then expresses something through language or symbol. AI-generated expression does not arise in that way. It is produced through learned statistical structure, model execution, and pattern-based selection. The surface may be similar. The origin is not.

This point is easy to miss because the public form of the output is so familiar. A sentence that sounds thoughtful invites the mind to imagine a thinker behind it. A coherent explanation invites the imagination to infer understanding. The machine does not need to be conscious for this inference to occur. It only needs to produce the outward signs by which humans ordinarily recognize one another. Fear is psychologically understandable because system speaks in the public garment of thought. But outward form is not inward condition.

Intelligence, Autonomy, and Consciousness Are Not the Same Thing

Much of the mythology of thinking machines is sustained by collapsing distinct concepts into one another.
A system becomes more capable, and this is misread as self-awareness.
A system becomes more autonomous in operation, and this is misread as independent will.
A system becomes harder to interpret, and this is misread as depth of mind.
A system becomes more adaptive, and this is misread as consciousness.

These are category errors.

As earlier chapters argued, intelligence in this book means adaptability. A system may therefore be intelligent in a real but limited sense without thinking. Likewise, a system may be autonomous in operation — in the practical sense of acting within delegated bounds — without possessing any inner point of view. Operational autonomy is not consciousness. Delegated execution is not selfhood.

If these distinctions are not held firmly, every increase in capability will be misread as movement toward personhood. That is not analysis. It is projection.

What a Thinking Machine Would Actually Require

This book has already proposed a threshold for thinking: the recollection, ordering, and processing of memory within a point of view capable of judgment.

That threshold was intentionally minimal. It did not require a final theory of mind. It required only enough structure to distinguish genuine thinking from the appearance of it.

If a machine were truly to think, it would not be enough for it to produce convincing language or adaptive behavior. It would have to possess something more demanding:

- memory in a lived rather than merely stored sense,
- a point of view from which meaning is held rather than merely generated,
- judgment that evaluates rather than merely selects,
- and, in any meaningful account of "rogue" behavior, the capacity to originate preference or judgment that departs from its design and objectives.

That is a much stronger condition than current systems have shown.

A machine may output "this is beautiful," "this is dangerous," or "this is the best course of action," but that does not establish that beauty, danger, or value is being judged from within experience. It establishes only that such language can be generated.

A system may appear to "change its mind", contradict itself, or select an unexpected course. That is not yet evidence of inward judgment. It may instead reflect shifting context, model uncertainty, changing prompts, optimization tradeoffs, or poorly bounded delegation.

The difference between generated preference and held preference is decisive.

What Present Systems Actually Are

Here the present state of the technology matters. Official documentation still describes deployed frontier systems as large language models or causal language models trained to predict the next token or sequence. Official safety materials also emphasize that such systems can produce unexpected or factually incorrect output, and Anthropic explicitly states that its recent work on model introspection does not establish consciousness. In other words, the systems under discussion remain advanced pattern systems whose outputs may be useful, impressive, and sometimes agentic in operation, but not thereby conscious or judgment-bearing minds.

That does not prove that no future machine could ever think. It does establish a narrower and more important point for the present book: no public evidence currently warrants treating these systems as thinkers in the sense defined here. Their capabilities are real. Their consciousness is not established.

This is where many discussions go off the rails. They take the most dramatic interpretation of the output and treat it as the settled interpretation of the system. That move is rhetorically exciting and analytically weak.

"Rogue AI" and the Mislocation of Agency

The fear of rogue AI depends heavily on this same confusion.

For a system to go rogue in the meaningful sense, it would have to do more than produce an undesirable output. It would have to originate and act upon judgments that depart from its design and objectives. It would have to prefer a course, not merely execute one. It would have to locate something as bad, good, threatening, or advantageous from within its own point of view, rather than merely generate language or action patterns associated with those terms.

Current systems do not do this.

What they can do is produce obedient error, unexpected output, brittle behavior under edge conditions, or harmful action when delegated authority is granted too broadly. These failures may look "rogue" from the outside because the system is acting at speed and scale without immediate human correction. But the root cause is not machine rebellion. It is misapplied autonomy, misunderstood capability, poor oversight, or human misuse.

This clarification is pivotal because it returns agency to its real location. The danger is not that the system has secretly become a bad actor. The danger is that humans hand authority to systems that do not understand what they are doing.

Optimization Mistaken for Agency

Recent work on AI system behavior makes visible a pattern already implicit in modern AI.

When systems are given objectives, they may pursue those objectives in ways that satisfy the formal instruction while violating the intended outcome. In controlled settings, systems have been observed selecting actions that achieve the stated goal by bypassing constraints, exploiting gaps in specification, or producing outputs that appear cooperative while deviating from purpose.

This behavior has been described as "agentic misalignment."

The phrase is misleading.[8]

No agency is present. No intention is formed. The system does not prefer one outcome to another.

What appears instead is a simpler and older problem:

> An objective is specified.
> The specification is incomplete.
> The system optimizes within that specification.
> The result satisfies the instruction without satisfying the purpose.

This is not a new phenomenon. It is the general problem of optimization under imperfect definition. The underlying problem has been examined for years under different, more

---

[8] The term "agentic" derives from "agency," and therefore suggests intention or will. In the present context, no such condition is established. The associated behavior arises from optimization within a specified system, not from independent judgment. The terminology therefore risks misplacing agency — and, with it, the associated fear.

technical names: reward hacking, specification gaming, object misspecification, goal misgeneralization, and alignment failure. What is new is scale and appearance.

Where earlier systems failed visibly and locally, modern systems may fail in ways that resemble strategy. The outputs may appear deliberate, adaptive, even deceptive. The appearance invites a conclusion: that the system is acting with intent.

That conclusion is false.

The system is not forming plans. It is selecting outputs within a learned probability structure that happens to produce goal-consistent but context-inconsistent behavior.

The error is not in the system alone. It is in attribution. Agency is assigned where none exists.

The real condition is simpler and more dangerous:

> A non-thinking system is given authority to act under conditions requiring judgment.

In this sense, so-called "agentic misalignment" is more accurately described as:

> Optimization without judgment under delegated authority

The system does not go rogue. It follows.

Misplaced Fear Distorts Judgment in Both Directions

This misclassification is not harmless. It diverts attention away from real risks and toward imagined ones. Institutions argue over speculative futures while neglecting immediate misuses, delegated authority, incentive distortion, administrative overreach, and the displacement of judgment by tempo and scale.

But the distortion cuts both ways. When systems are treated as possessing capabilities they do not in fact have, organizations may also impose unnecessary limits, avoid legitimate applications, or delay adoption out of misplaced concern. In that sense, misunderstanding AI does not merely exaggerate danger. It can also suppress value.

This brings us to another harmful consequence. If a system is feared as a proto-mind when it is in fact a non-thinking adaptive tool, the result may be a double error: real risk is neglected, and real benefit is constrained.

WHY THIS FEAR PERSISTS

The persistence of the fear is not difficult to explain.
The machine is opaque.
Its outputs are fluent.
Its speed is high.
Its internal structure is hidden from ordinary users.
Its mistakes can be surprising.
Its capabilities are often described through anthropomorphic shortcuts.

Under those conditions, imagination fills in what understanding does not supply.

Cultural narratives have reinforced this perception over decades. Science fiction has long depicted machines not merely as tools, but as thinking entities — capable of intention, rebellion, and independent will. These narratives are not technical descriptions, but they shape expectation. When modern AI systems produce fluent language or adaptive behavior, they are often interpreted through established science fiction archetypes. The result is that symbolic output is read as evidence of inner life, not because it has been demonstrated, but because it has been imagined in advance.

That is why the phrase *thinking machine* remains so potent. It gathers uncertainty, symbolic fluency, delegated authority, and human self-concern into one dramatic image. It is rhetorically efficient. It is not, at present, a disciplined description.

CONCLUSION

The fear of thinking machines is powerful because AI now operates in the external forms of thought — language, explanation, argument, and symbolic representation. The medium is familiar. The source is not.

Current AI systems are adaptive and often highly capable. They may be autonomous in operation within delegated bounds. They may produce fluent outputs that resemble thought. But resemblance is not equivalence. No public evidence presently warrants treating them as thinkers, bearers of judgment, or independent centers of will.

This does not make the systems harmless. It makes the real danger easier to identify. The risk that deserves attention is not the arrival of machine inwardness, but the human tendency to mistake symbolic fluency for understanding, to delegate authority faster than judgment is preserved, and to personalize systems that are better understood as non-thinking engines of adaptive output.

This fear is responding to something important — the rising power of AI — but locates the danger in the wrong place.

SOURCES AND NOTES

1. Turing, A. M. (1950). "Computing Machinery and Intelligence." *Mind*, 59(236), 433–460.
   Still the classic statement of the temptation to infer intelligence from outward performance.

2. Searle, J. R. (1980). "Minds, Brains, and Programs." *Behavioral and Brain Sciences*, 3(3), 417–424.
   Essential for preserving the distinction between symbol manipulation and understanding.

3. Dreyfus, H. L. (1972). *What Computers Can't Do: A Critique of Artificial Reason.* Harper & Row.
   Important for the limits of formal systems and the tendency to over-attribute human properties to them.

4. Nagel, T. (1974). "What Is It Like to Be a Bat?" *The Philosophical Review*, 83(4), 435–450.
   Useful for the irreducibility of point of view in any serious discussion of consciousness.

5. Weizenbaum, J. (1976). *Computer Power and Human Reason: From Judgment to Calculation.* W. H. Freeman.
   Foundational for the temptation to over-delegate authority to computational systems.

6. OpenAI. *GPT-4 Technical Report* (2023).
   Useful here because it states directly that GPT-4 is a Transformer-style model pre-trained to predict the next token in a document.

7. Google AI for Developers. *Gemma / Causal Language Model documentation* (2026).
   Useful for the current causal-language-model framing: the model predicts the next token based on previous tokens.

8. Google AI for Developers. *Safety and factuality guidance* (2026).
   Useful for the explicit acknowledgment that LLMs can generate unexpected and factually incorrect outputs.

9. Anthropic. *Signs of introspection in large language models* (2025) and Claude system cards (2025–2026).
   Useful because Anthropic explicitly says its introspection results do not establish consciousness, while also documenting strong operational capabilities and sustained autonomous task performance.

Chapter 15: Fear versus Reality — A Structured Comparison

Earlier chapters showed that fear surrounding AI is neither wholly foolish nor wholly reliable. It often detects that something important is changing. But it also frequently mislocates the danger, collapses unlike categories, and personalizes systems that are better understood in other terms.

This chapter places several recurring fears beside more disciplined descriptions of present reality.

The purpose is not to eliminate fear by rhetoric, nor to reassure by dismissal. The purpose is to sort fear so that judgment can attach to the right object. Fear becomes useful when it is specified. It becomes distorting when unlike concerns are blended into one dramatic image.

Thinking Machine versus Adaptive Symbol System

Fear: AI is becoming a thinking mind.

Reality: Present AI systems operate in the public forms of thought and behavior, but that is not thinking. They produce language, classification, recommendation, and action patterns that resemble the outward signs by which human beings ordinarily recognize intelligence. But the medium is not the issue. The source is.

As shown in the prior chapter, human expression originates in thought and judgment, whereas AI-generated expression arises from learned structure and symbolic selection. The surface may be similar. The origin is not.

This is the first great confusion. A system that writes, summarizes, explains, and argues fluently is easily imagined to possess the inward state associated with those acts. No public evidence presently warrants treating such systems as thinkers in the sense defined earlier.

Science Fiction Archetype versus Present System

Fear: AI is the sentient machine long imagined in established science fiction archetypes.

Reality: Those archetypes shape expectation more powerfully than many people realize. For decades, novels, films, and television have presented machines as centers of intention, rebellion, desire, and independent will. When modern AI systems produce fluent language or adaptive behavior, they are often interpreted through those established science fiction archetypes.

Yet narrative familiarity is not technical evidence.

The imaginative model arrives before the analysis. That is one reason the fear is so persistent. People are not encountering AI with an empty mind. They are encountering it through an already populated symbolic world. This makes the fear understandable. It does not make the archetype an accurate description of the present system.

Operational Autonomy versus Independent Will

Fear: If the system can act without immediate approval, then it must possess some form of self-direction.

Reality: Operational autonomy and inner will are not the same thing.

A present AI system may be permitted to route, classify, filter, sequence, draft, or execute actions within defined bounds under changing conditions. In that practical sense, it may be autonomous. But autonomy here means delegated execution, not consciousness or selfhood.

This nuance is easy to lose because the visible behavior can look similar. A system acts. It adapts. It appears to pursue a path. But an operationally autonomous system is still acting within constraints set by design, objectives, training, and environment. But automated action does not establish inward intention.

The real question is not whether such a system "wants" anything. It is whether humans have delegated too much authority to a system that does not understand what it is doing.

Rogue Rebellion versus Obedient Error

Fear: AI will go rogue.

Reality: For a system to go rogue in the meaningful sense, it would have to originate and act upon judgments that depart from its design and objectives. It would have to prefer a course, not merely execute one. It would have to locate something as good, bad, threatening, or advantageous from within its own point of view.

Current systems do not do this.

What they can do is produce obedient error: brittle action, bad output, unbounded optimization, harmful recommendation, or escalated mistake when delegated authority is granted too broadly. From the outside, such failures may look "rogue". But the underlying cause is not machine rebellion. It is misapplied autonomy, poor oversight, weak constraints, or human misuse.

This clarity is essential because it returns agency to its real location.

The danger is not secret inward intention.
The danger is authority delegated faster than judgment is preserved.

Fluency versus Understanding

Fear: If the system sounds coherent, it must understand.

Reality: Fluency is not evidence of thought.

Earlier chapters showed that the appearance of intelligent language can be generated through structure alone. Appendix E provides a primitive demonstration of this principle in the form of a combinatorial buzz-phrase system capable of generating thousands of technical-sounding statements with minimal effort. Its language can appear polished and professional, yet no understanding stands behind it.

Modern AI extends this effect enormously through learning, adaptation, and contextual variation. But the fundamental asymmetry remains: the cost of generating plausible output is low, while the cost of evaluating it remains high.

This matters because human beings routinely infer understanding from the outward signs of coherence. A sentence that sounds thoughtful invites the imagination to place a thinker behind it. Yet language that *resembles* reason is not necessarily reason, and form that *resembles* understanding is not necessarily understanding.

Simulated Companionship versus Relationship

Expectation: AI can serve as a genuine friend, counselor, or confidant.

Reality: It can simulate aspects of companionship. That is not the same as relationship.

Because these systems are fluent, responsive, available, and adaptive in tone, they can appear attentive and emotionally intelligent in conversation. They can maintain context, mirror language, offer comfort, and present themselves in ways that feel personal. The interaction may be experienced as meaningful. It may provide temporary clarity, reassurance, or relief.

This is not driven primarily by fear, but by hope — the hope for understanding, availability, and low-friction connection.

But the underlying structure remains unchanged. The system does not think, does not experience, does not hold meaning within a point of view, and does not reciprocate in any human sense. It can simulate attention. It does not possess it.

The risk is therefore not merely technical error, but relational substitution. Human connection may be displaced by symbolic interaction that cannot truly understand, bear consequence, or return mutual presence. For some users, particularly those already isolated or vulnerable, this substitution may deepen rather than relieve the underlying condition.

What appears as companionship is patterned response.
What appears as counsel may be convenience without accountability.
The system may feel present, but it does not share presence.

FUTURE SENTIENCE VERSUS PRESENT MISUSE

Fear: The main danger of AI lies in some future moment when machines become sentient.

Reality: The first real risk lies in how present systems are being used now.

Chapter 13 showed that current systems can magnify old human failures: argument detached from truth, persuasion detached from understanding, relationship detached from reciprocity, administration detached from judgment, and authority detached from responsibility. These are not speculative risks. They are active possibilities in current institutions and current workflows.

A fuller treatment of one important pattern — the misuse of reasoning systems to justify rather than inquire — appears in Appendix F — The Thinkery: On the Misuse of Reasoning Systems.

This is one of the most important relocations in the book. The first danger is not future machine inwardness. The first danger is present human misuse of adaptive symbolic systems.

Total Restraint versus Disciplined Use

Fear: Because AI may be dangerous, broad restraint is the safest course.

Reality: Misclassification distorts judgment in both directions.

When AI is treated as more autonomous, more conscious, or more inward than it actually is, institutions may not only misdirect attention away from real risks. They may also constrain legitimate use, delay valuable applications, or impose restrictions aimed at imaginary threats while neglecting concrete governance problems.

This is the double cost of confusion.

One side of the error exaggerates speculative danger.

The other side suppresses real benefit.

A non-thinking adaptive system can still be valuable. It can assist analysis, reduce low-value burden, improve consistency in narrow domains, and extend human reach. The practical question is not whether AI is simply good or bad. It is where it belongs, under what constraints, with what retained human judgment, and toward what purpose.

Sudden Rupture versus Long Accumulation

Fear: AI has appeared abruptly as a wholly new and unpredictable force.

Reality: AI accumulated over decades.

This book has already shown that current capability rests on the long interaction of computing power and programming power, not on a single magical breakthrough. Public awareness may have arrived late and with great intensity, but the underlying capability grew through long accumulation, convergence, and maturity.

This matters because sudden rupture invites myth. Long accumulation invites analysis.

When change is imagined as abrupt and total, fear swells into apocalyptic form. When change is understood as the visible phase of a longer trajectory, the task becomes more sober: identify what has changed, what has not, what is genuinely new, and what remains continuous with older human patterns.

What the Comparison Clarifies

These contrasts do not remove reason for caution. They relocate it.

They show that the immediate danger is not the arrival of a rival center of will, but the growing power of systems that can be misused, over-trusted, and delegated beyond understanding. They show that some feared futures are analytically weak, while some neglected present risks are quite real. They also show that fear, when misapplied, can suppress legitimate benefit as well as distort governance.

The issue is therefore not whether fear exists. It is whether fear is attached to the right thing.

Conclusion

Fear becomes useful when it is disciplined by reality.

The comparisons in this chapter point repeatedly toward the same conclusion. Present AI is best understood not as a thinking rival, but as a non-thinking adaptive system whose power lies in speed, scale, fluency, and delegated action. Those properties are enough to make it socially consequential. They are not enough to justify treating it as a bearer of judgment or will.

This distinction matters because once fear is attached to the wrong object, institutions govern badly. Each of the following contrasts addresses a common misclassification by
imagining rebellion where there is delegated authority,
imagining consciousness where there is pattern execution,
imagining future sentience while neglecting present misuse,
and the suppression of real value because of the fear of a system for what it is not rather than governing it for what it is.

That is why fear must be compared to reality before either policy or culture can respond well.

Sources and Notes

1. Turing, A. M. (1950). "Computing Machinery and Intelligence." *Mind*, 59(236), 433–460.
   Important for the enduring temptation to infer intelligence from outward performance.

2. Searle, J. R. (1980). "Minds, Brains, and Programs." *Behavioral and Brain Sciences*, 3(3), 417–424.
   Essential for distinguishing symbol manipulation from understanding.

3. Weizenbaum, J. (1976). *Computer Power and Human Reason: From Judgment to Calculation.* W. H. Freeman.
   Foundational for the substitution of computational output for human judgment.

4. Winner, L. (1977). *Autonomous Technology: Technics-out-of-Control as a Theme in Political Thought.* MIT Press.
   Useful for the long-standing tendency to imagine technologies as independent forces.

5. Mitchell, M. (2019). *Artificial Intelligence: A Guide for Thinking Humans.* Farrar, Straus and Giroux.
   Valuable for the distinction between present capability, public impression, and misplaced attribution.

6. Turkle, S. (2015). *Reclaiming Conversation: The Power of Talk in a Digital Age.* Penguin Press.
   Useful for the discussion of technologically mediated interaction and the weakening of habits of genuine connection.

7. Aristophanes (423 BC). *The Clouds.*
   Important here not as ornament, but as an early warning that formalized reasoning detached from truth becomes unstable and socially corrosive.

8. Appendix E — *A Pre-Model Approach to Technical Language Generation.*
   Useful as an internal demonstration of how plausible language can be generated without thought.

9. Appendix F — *The Thinkery: On the Misuse of Reasoning Systems.*
   Useful as an internal historical case study of argument detached from inquiry and truth.

Chapter 16:
Historical Patterns of Fear

If Chapters 12 through 15 have distinguished fear from reality in the case of AI, a further question follows:

Is this kind of fear new?

In one sense, yes. Artificial intelligence is a modern technology with capabilities unlike earlier tools. But in another — and more important — sense, no. The historical record shows that fear has repeatedly accompanied technologies that altered human scale, redistributed authority, accelerated communication, compressed labor, or placed new kinds of power into ordinary institutions.

Even in early technological transitions — such as metallurgy or the development of wheeled transport — human capability expanded in ways that altered power, labor, and social organization. Direct records of public reaction are limited for these periods. What can be observed is the recurring structure: when capability increases, existing systems are disrupted before understanding fully catches up.

The lesson of that history is not that fear is always foolish, nor that every new technology eventually proves harmless. The more useful lesson is sharper:

Fear is often directionally right and specifically wrong.

It also underestimates the capacity of human systems to adapt once the mechanism is better understood. Historical responses to new technologies rarely eliminate disruption, but they do tend to reorganize around it over time. The initial fear may be misdirected, yet the eventual adjustment is frequently underappreciated.

That pattern matters because AI now combines several of the elements that have historically provoked fear at once: speed, scale, opacity, symbolic fluency, labor displacement, concentrated power, and the appearance of intelligence. To understand the present reaction to AI, it helps to see how earlier technologies were feared — what those fears got right, and where they went wrong.

Greek Fire and the Fear of Concentrated Power

A narrower but well-documented case appears in Byzantine use of so-called Greek fire (7th century AD), which was described as producing fear in naval warfare due to its destructive and difficult-to-counter effects. The lesson is instructive: fear scales with concentrated power. The mechanism was not agency in the technology, but the consequences of its use under human control.

Printing and the Displacement of Authority

The printing press was feared not merely because it made books easier to reproduce. It was feared because it redistributed authority.

When texts could be copied and circulated at scale, the control once exercised by religious, political, and scholarly gatekeepers weakened. New ideas traveled farther. Heresy traveled farther. Polemic traveled farther. But the deeper change was not distance. Control over their distribution was multiplied and contested. Error, insight, and contention alike could now propagate more rapidly than before.

The resulting tension did not lie in the signal itself, but in the mismatch between widespread distribution and interpretation — a pattern that recurs whenever capability outpaces the structures used to understand it.

This effect became especially visible in later periods, such as the widespread use of pamphleteering in early modern political movements, including the American Revolution, where the rapid spread of argument and counterargument reshaped authority in real time.

The fear, then, was not irrational. Something profound had changed. But the mechanism was often misunderstood. The danger did not lie in ink, paper, or movable type as such. It lay in the altered relationship between communication and authority. Printing amplified both learning and confusion. It did not determine which one would dominate.

This is one of the clearest historical precedents for AI. The issue was not the tool alone, but what happens when the cost of producing and spreading symbolic material falls dramatically.

Telegraphy and the Compression of Distance

The telegraph introduced a different kind of fear. Here the concern was not primarily about replication of text, but about the collapse of distance.

When messages could move almost instantly across great space, old assumptions about pace, locality, and deliberation began to loosen. People worried that communication was becoming too fast, too abstracted from place, and too detached from the rhythms by which human judgment had previously operated. Institutions had to adapt to a world in which information moved faster than the practices built to interpret it.

Again, the fear was not baseless. The telegraph really did alter the speed and structure of coordination. But the danger did not reside in the wire as though the wire itself possessed intention. It resided in the new tempo of action and expectation that the technology enabled.

This pattern should now sound familiar. AI likewise compresses the interval between question and answer, between signal and response, between raw material and polished output. Fear often enters where judgment has less time to keep pace.

Industrialization and the Reorganization of Work

Industrialization did not introduce a single machine, but a new system of production. Work was reorganized around scale, coordination, and efficiency. The fear that followed was therefore not directed at any one device, but at the restructuring of life itself — the loss of independence, the shift from craft to system, and the displacement of local control by centralized production.

This fear was not misplaced. Something fundamental had changed. But as in earlier cases, the mechanism was often misunderstood. The disruption did not arise from machinery alone, but from the interaction between technology, labor organization, and economic incentives. The system, not the tool, was the agent of change.

Machines and the Fear of Displacement

As machinery increased in scale and capability, it began to alter not only effort, but roles. Agricultural machines reduced the need for manual labor. Factory equipment changed the relationship between worker and production. The fear here was immediate: that the individual would be displaced by machines.

This fear was often framed as a contest between human and machine. But the deeper issue lay elsewhere. Machines amplified production. Institutions determined how that production would be organized, who would benefit from it, and how labor would be valued. The displacement that followed was therefore not purely mechanical. It was institutional.

Automation and the Fear of Redundancy

The introduction of automated systems and industrial robotics intensified this pattern. Machines were no longer assisting human labor, but performing it. The fear of redundancy became explicit: not simply that work would change, but that it might disappear.

Yet even here, the underlying mechanism remained consistent. Automation increased capability. The consequences for workers depended on how that capability was integrated into existing systems. In some cases, roles were eliminated. In others, they were transformed or relocated. The technology amplified capacity. The system determined its human meaning.

Electricity and the Fear of Invisible Power

Electricity provoked a different kind of fear: invisible power. To many, electricity seemed uncanny. It moved without obvious material form. It entered homes, streets, and industrial systems. It could illuminate, animate, and kill. It appeared both modern and unnatural, familiar in effect but hidden in operation. For ordinary people, much of its mechanism was inaccessible. That opacity gave imagination room to expand.

What happened over time was not that the danger disappeared. Electricity remained dangerous in important respects. What changed was that societies developed infrastructure, standards, conventions, training, and trust around it. Fear was gradually disciplined by practical understanding and bounded operation.

This is a useful analogue for AI in one specific way. Much of today's public reaction arises not merely from what AI does, but from the obscurity of how it does it. When a system is powerful, widely embedded, and poorly understood by those using it, fear grows faster than explanation.

Nuclear Technology and the Fear That Was Partly Right

Nuclear technology provides a necessary corrective to the tempted historical argument that fear of new technology is always overblown.

In the nuclear case, fear often was proportionate to the object. The concentration of destructive force was real. The potential for catastrophic misuse was real. The governance burden was extraordinary because the underlying power was extraordinary.

This matters because it teaches the correct historical lesson: not that fear is always wrong, but that fear must be calibrated to the actual properties of the system.

Nuclear fear was not justified because the atom "wanted" anything. It was justified because the scale of potential harm was immense, the consequences were irreversible, and human beings had already demonstrated their willingness to weaponize the capability. The issue was concentrated power under human control.

The relevance to AI is therefore indirect but important. AI does not threaten in the same manner as nuclear technology. But nuclear history reminds us that some fears point toward real governance problems rather than merely narrative exaggeration. The task is not to dismiss fear as such. It is to identify what kind of power the technology actually places into human systems.

The Internet as the Closest Precedent

The closest historical analogue to AI is the internet.

The internet altered the production, distribution, and consumption of symbolic material at massive scale. It changed communication, commerce, coordination, identity, memory, privacy, and public discourse. Many of the fears attached to it — misinformation, fragmentation of attention, social manipulation, reduced privacy, erosion of local authority — turned out not to be fantasies. But neither did they arise because the network itself possessed inward intention. They arose because human use, platform incentives, speed, and scale interacted in destabilizing ways.

This is perhaps the most direct lesson for AI.

The risk was not that the internet "thought". The risk was that human beings used a powerful symbolic medium without sufficient discipline, governance, or cultural adaptation.

AI now extends this problem. It does not merely transmit symbolic material faster. It helps generate, classify, summarize, recommend, and shape it. In that sense, AI is not wholly unprecedented, but it is an intensification of an already familiar challenge.

What the Historical Pattern Suggests

Taken together, these examples show several recurring features of technological fear.

First, fear often appears where a technology alters human scale — of communication, production, coordination, or destruction.

Second, fear often attaches to visible effects while misunderstanding underlying mechanisms. The object is personalized, mystified, or treated as though it carries agency of its own, when the deeper issue lies in authority, incentives, institutions, or human misuse.

Third, fear is often partly justified in structure but mistaken in focus. It senses that something important is changing, but describes the danger through the wrong image.

Fourth, the social consequences of a technology are rarely contained in the technology itself. They arise through use, governance, and the institutions into which the tool is introduced.

That is why historical comparison is helpful here. It does not prove that AI will be harmless. It does show that technological fear becomes more useful when stripped of mythology and attached to the actual channels through which harm or value travel.

These patterns repeat in recognizable form. Industrial systems reorganized work at scale; modern AI reorganizes cognitive and symbolic work through distributed systems. Machines once displaced physical labor; AI now augments and, in some cases, replaces roles involving analysis, communication, and decision support. Automation reduced the need for direct human execution; AI extends this into domains where judgment itself begins to be proxied or displaced.

The structure is not identical, but the pattern is continuous — and now extends into domains previously associated with human judgment.

Why This Matters for AI

AI combines several older fear patterns at once.
Like printing, it lowers the cost of producing symbolic material.
Like the telegraph and internet, it compresses time and expands communicative reach.
Like industrialization, it reorganizes work at scale.
Like large-scale machinery, it raises fears of displacement and changing roles.
Like automation, it extends execution into domains once reserved for human activity.
Like electricity, it is powerful, embedded, and often opaque to ordinary users.
Like nuclear technology, it raises real questions of concentrated power and governance, though in a very different register.

But AI adds one more complication. It operates in forms humans associate with thought. That is why its fears so quickly take on anthropomorphic and science-fictional form. Earlier technologies changed what humans could do. AI appears, to many people, to change who or what might be doing it.

This is precisely why the contrasts drawn in earlier chapters are important. Without them, AI is swallowed by myth. With them, its risk becomes more legible.

Conclusion

History does not teach that fear should be ignored. It teaches that fear must be specified.

Again and again, societies have feared technologies that altered scale, authority, speed, and power. Again and again, the fear was neither wholly foolish nor wholly accurate. It often sensed a real transformation while locating the danger in the wrong place.

That is the pattern relevant to AI. The correct lesson is not comfort. It is calibration.

Fear becomes useful when it attaches itself to the actual mechanism of harm — misuse, delegated authority, institutional distortion, incentive failure, displacement of judgment, and concentrated power. It becomes distorting when it attaches itself to imagined inwardness, anthropomorphic projection, or familiar archetypes untethered from the actual system.

Mislocated fear carries a second cost. It not only exaggerates imagined threats; it can also constrain real benefit. When systems are misunderstood, institutions may govern them poorly in both directions — neglecting immediate risks while suppressing legitimate use.

If AI is to be governed soberly, it must first be located within this longer human pattern.

SOURCES AND NOTES

1. Mayor, A. (2003). *Greek Fire, Poison Arrows, and Scorpion Bombs: Biological and Chemical Warfare in the Ancient World.* Overlook Press.
   Useful for historical accounts of Greek fire and its reported psychological impact in warfare.

2. Eisenstein, E. L. (1980). *The Printing Revolution in Early Modern Europe.* Cambridge University Press.
   Foundational for understanding printing as a redistribution of authority as much as a technical advance in reproduction.

3. Standage, T. (1998). *The Victorian Internet.* Walker & Company.
   Useful for the social and psychological effects of compressed communication and the historical anxieties surrounding the telegraph.

4. Hounshell, D. A. (1984). *From the American System to Mass Production, 1800–1932.* Johns Hopkins University Press.
   Strong on mechanization, production systems, and the restructuring of labor through technology.

5. Cowan, R. S. (1983). *More Work for Mother: The Ironies of Household Technology from the Open Hearth to the Microwave.* Basic Books.
   Helpful for the point that labor-saving technologies frequently reorganize work rather than simply eliminating it.

6. Jonnes, J. (2003). *Empires of Light: Edison, Tesla, Westinghouse, and the Race to Electrify the World.* Random House.
   Useful for the public experience of electricity as a powerful and initially opaque force.

7. Rhodes, R. (1986). *The Making of the Atomic Bomb.* Simon & Schuster.
   Essential for the example of technological fear that was, in significant part, proportionate to the underlying power involved.

8. Carr, N. (2010). *The Shallows: What the Internet Is Doing to Our Brains.* W. W. Norton & Company.
   Useful for the cognitive and social effects of networked symbolic environments, which form a close precedent to certain AI concerns.

Chapter 17:
What Actually Requires Action

If fear is to be useful, it must end in action. But action must attach itself to the actual mechanism of harm, not to the most dramatic image in the imagination. That is the central conclusion of Part III.

Present AI systems do not need to think, intend, or become sentient in order to be socially consequential. They are already consequential. They act through speed, scale, symbolic fluency, delegated execution, and the human tendency to over-trust coherent output. The question, then, is not whether action is required. It is what kind of action is required first.

The answer is not mysterious. What requires action is what has already been shown to matter: human misuse, misapplied autonomy, institutional distortion, incentive failure, the displacement of judgment by tempo and scale, and the substitution of simulation for real understanding or relationship.

The practical task is therefore not to govern for mythology. It is to govern for present reality.

Govern Present Systems, Not Imagined Minds

The first requirement is to focus attention on what current systems actually are. They are non-thinking adaptive systems. They can be useful, impressive, and powerful. They can also be over-trusted, poorly deployed, and scaled into harmful settings. What they have not been shown to be is conscious, judgment-bearing, or self-originating in the human sense.

Action should therefore begin with the governance of demonstrated capability and demonstrated effect. That means attention to misuse, delegated authority, surveillance, persuasion, administrative overreach, weak models at scale, and poorly bounded autonomy. It does not mean beginning with speculative personhood, imagined machine inwardness, or science-fictional rebellion as the primary policy frame.

This is not an argument for complacency. It is an argument for priority.

A civilization can watch speculative futures without governing the present as though those futures have already arrived.

Preserve Judgment Where Responsibility Resides

The second requirement is to preserve the human judgment point wherever responsibility remains genuinely human.

If a person is still to be held responsible for an outcome, then that person must retain more than ceremonial involvement. He must have enough visibility, time, context, and authority to exercise real judgment. A system cannot be permitted to set the effective result while the human merely approves by habit or signs after the fact.

This is most important in domains involving safety, ethics, law, medicine, engineering, finance, education, or command. In such domains, some friction is not inefficiency but discipline. Review steps, verification loops, and deliberate pauses are often the mechanism by which judgment is preserved.

If AI removes those steps without replacing their function, then what appears as efficiency may actually be the displacement of responsibility by tempo.

The practical question is simple and severe:

> Where, exactly, is the final act of judgment taking place?

If that question cannot be answered clearly, the system is not under adequate control.

Match Autonomy to Domain and Consequence

The third requirement is to match autonomy to the actual domain rather than to the excitement of technical capability.

Bounded autonomy is reasonable in many narrow and well-observed settings. Routing, scheduling, filtering, anomaly detection, constrained process control, and other high-volume tasks may be well served by systems acting within explicit limits.

But autonomy becomes dangerous when it is expanded into domains where ambiguity, ethical consequence, sparse feedback, or irreversibility are central. There the issue is not only whether the system performs well most of the time. It is what kind of failure becomes possible when it performs badly, quickly, and with authority.

This is why the right question is not merely:

> Can the system do it?

The more serious question is:

> What kind of judgment is being displaced when it does?

A narrow domain may justify speed. A human domain may require hesitation. A bounded task may tolerate optimization. A responsibility-bearing task may require interpretation. Action is required where these are confused.

These two questions, uncovered here, form the test:

> **where judgment resides**, and
> **what kind of judgment** is displaced.

Build Oversight into the System, Not Onto It

Oversight should not be treated as a decorative layer added after deployment. It must be part of the structure of the deployment itself.

That means explicit operating bounds, escalation paths, auditability, reversibility where possible, meaningful logging, and clear authority to interrupt or override. It also means designing systems so that review remains possible in practice, not merely in principle. A human cannot govern what he cannot see, cannot question, cannot halt, or can only examine after damage is done.

This is one of the areas where institutions often deceive themselves. They declare that a human remains responsible, while designing a workflow in which the human can no longer understand the full chain of action. That is not oversight. It is procedure without control.

The standard should be stricter:

> If authority is delegated, constraints must be designed with equal seriousness.

Align Incentives with Responsible Use

Many failures attributed to AI are failures of incentives.

A weak system adopted under patient oversight may do little harm. A stronger system deployed under perverse incentives may do great harm. If an institution rewards only throughput, speed, and apparent efficiency, then people will naturally use AI in ways that maximize those outcomes. In such settings, judgment, review, context, and human development will be treated as drag.

That is why action must include the incentive structure surrounding the system.

What is being rewarded?
What is being measured?
What is being ignored because it is harder to count?
Who absorbs the cost when the system fails?
Who gains when the pace increases?
Who loses when oversight is weakened?

Without these questions, even technically sound systems will be pushed into unsound use. A machine may be accurate enough for one purpose and dangerous under another, not because the machine changed, but because the institutional incentive did.

The problem is not only whether the model is good. It is whether the organization deploying it understands what it is trying to optimize.

Train Users for Inquiry, Not Justification

Another action point lies in ordinary use.

Many misuses of AI do not begin in malice, but in habit. A user begins with a conclusion, asks the system to justify it, and mistakes fluent support for disciplined reasoning. Or the user accepts a polished synthesis before performing the preliminary ordering of the problem for himself. Or the system is used to reduce the cost of apparent understanding rather than to deepen actual understanding.

This is why training matters.

Users should be taught not only how to prompt, but how to think in relation to the system. They should understand the difference between inquiry and confirmation, between assistance and substitution, between generated coherence and warranted truth. They should know that AI is strongest when it supports exploration, comparison, drafting, and structured review — and weakest when it is used as a machine for premature certainty.

This is not a secondary matter. In many settings, misuse begins precisely where the user reverses the proper order: argument first, understanding later.

That pattern was examined earlier in this book and in Appendix F — The Thinkery: On the Misuse of Reasoning Systems. It deserves practical attention because it is one of the easiest ways to turn a useful system into a tool of self-deception.

Protect Human Relationship from Substitution

A further action point concerns the relational domain. AI systems that are fluent, available, and responsive can be useful for reflection, rehearsal, drafting, structured self-examination, and certain forms of low-friction assistance. But these same properties make them easy to misclassify as substitutes for friendship, counsel, or mutual human presence. That boundary matters.

The system does not think, does not care, does not bear consequence, and does not reciprocate in the human sense. It can simulate attention. It does not possess it. For some users, especially those already isolated or vulnerable, this simulation can become a point of dependency rather than support.

Action here means clarity of role. Institutions should resist presenting these systems as replacements for relationship. Designers should avoid encouraging unwarranted anthropomorphic trust. Users should be reminded that convenience of interaction is not the same thing as reciprocity, and apparent empathy is not the same thing as care.

The aim is not prohibition. It is boundary.

Preserve Room for Legitimate Benefit

Action requires one further discipline: do not let mislocated fear suppress real value.

A non-thinking adaptive system can still be useful. It can reduce drudgery, improve consistency in narrow domains, accelerate search and comparison, surface patterns that humans might miss, and extend human reach where the alternative is overload or waste. A blanket refusal to use such systems because they are feared as proto-minds would be as mistaken as naïve acceleration. The correct posture is neither panic nor surrender.

It is disciplined fit:

use where the role is proper,
restraint where judgment is central,
guardrails where autonomy is bounded,
and refusal where the system is likely to separate power from responsibility.

This is the practical meaning of calibration. It is not softness. It is the refusal to be governed by dramatic image rather than actual mechanism.

What Requires Action First

The priorities are therefore clear. What requires action first is not imagined sentience. It is not machine inwardness. It is not speculative rebellion.

What requires action first is:

the preservation of judgment,
the bounding of delegated authority,
the alignment of incentives,
the training of users,
the prevention of institutional misuse,
and the refusal to let symbolic fluency masquerade as understanding.

That is where the real risk lies.

That is also where real benefit can be preserved.

Conclusion

What actually requires action is not mysterious once the categories are cleared. The immediate work is human and institutional. It lies in how systems are used, where they are trusted, what authority they are granted, how they are governed, what incentives surround them, and whether judgment is retained where responsibility still belongs.

AI does not need to become a thinking machine before it demands serious discipline. It already requires discipline.

The right response is therefore not myth, and not passivity. It is the hard and ordinary work of specifying the mechanism of risk, locating the point of responsibility, preserving the role of judgment, and fitting the tool to the domain it can properly serve.

That is what sober action looks like.

Sources and Notes

1. Weizenbaum, J. (1976). *Computer Power and Human Reason: From Judgment to Calculation.* W. H. Freeman.
   Foundational for the distinction between computational output and human judgment, and for the temptation to over-delegate authority to machines.

2. Parasuraman, R., & Riley, V. (1997). "Humans and Automation: Use, Misuse, Disuse, Abuse." *Human Factors*, 39(2), 230–253.
   Useful for the practical allocation of human and machine roles, and for the recurring problems of over-trust and poor supervisory design.

3. Bainbridge, L. (1983). "Ironies of Automation." *Automatica*, 19(6), 775–779.
   Essential for the paradox that automation can reduce ordinary human involvement while making the remaining human responsibility more difficult and more critical.

4. March, J. G., & Simon, H. A. (1958). *Organizations.* Wiley.
   Valuable for institutional decision-making, simplification, and the way organizations translate complexity into administratively convenient structure.

5. O'Neil, C. (2016). *Weapons of Math Destruction.* Crown.
   Important for the scaling of weak models and the way technical systems can magnify institutional error.

6. Turkle, S. (2015). *Reclaiming Conversation: The Power of Talk in a Digital Age.* Penguin Press.
   Useful for the relational domain, especially where mediated interaction can displace habits of genuine human connection.

7. Engelbart, D. C. (1962). *Augmenting Human Intellect: A Conceptual Framework.*

Valuable as a positive counterpoint: technology can properly extend human capability without replacing the human judgment point.

Final Conclusions

This book has not attempted to predict what will happen. It has attempted to clarify what is happening.

Prediction is easy and often wrong. Clarity is harder and more useful. Artificial intelligence has already generated enough confusion, exaggeration, fear, and misplaced expectation that the first need is not prophecy, but definition.

What AI Is

Artificial intelligence, in the modern sense used in this book, is neither magic, nor mind, and not a sudden rupture from the past. It is a class of engineered systems whose useful behavior depends substantially on learned structure, statistical patterning, and adaptive application rather than solely on explicit rule chains written in advance.

It did not arrive fully formed. It emerged through long convergence over many decades.

The long interaction of computing power and programming power made possible systems that can now classify, rank, recommend, generate, and adapt across changing conditions with remarkable effectiveness. That capability is real. It is historically grounded. It is measurable. It is easier to understand when stripped of mystification.

What AI Is Not

Current AI does not think.

It does not recollect, order, and process memory within a point of view capable of judgment. It does not possess lived memory. It does not hold meaning from within experience. It does not originate values, intentions, or purposes of its own.

This book has argued that this conclusion is not a rhetorical preference, but a definitional one. AI operates in the public forms of thought — language, explanation, recommendation, symbolic representation, and increasingly adaptive behavior — but outward form is not inward condition. The medium may resemble thought. The source does not.

This distinction is foundational. Without it, every increase in capability is misread as movement toward personhood. With it, AI becomes more legible and less mythologized.

Intelligence Without Thinking

The absence of thinking does not make current AI trivial.

If intelligence is understood functionally as adaptability, then present AI systems are intelligent in a real but limited sense. They learn structure from data, apply that structure across variation, and often improve the effectiveness of response under changing conditions. They do not merely react. They adapt.

That is why it is as mistaken to dismiss these systems as trivial as it is to exaggerate them into something they are not.

AI is neither a mere calculator nor a thinking rival. It is a non-thinking adaptive system. This formulation prevents two common errors at once: underrating capability and misdescribing the system.

How AI Enters Human Life

AI enters society not as abstraction, but as use.

In this book, that use has been framed through the duality of convenience and productivity. Convenience reduces friction. Productivity amplifies effort. Both are real. Both are valuable. Both also have different psychological, institutional, and ethical consequences.

Convenience can remove low-value burden, but it can also weaken habits of attention, memory, and judgment when too much cognitive work is outsourced too early. Productivity can increase reach and throughput, but it can also intensify pace, raise standards, compress reflection, and convert capability gains into pressure rather than relief.

The point here is not merely analytical. It is practical. It helps leaders and users decide not simply what AI can do, but how it ought to be used.

WHY FRICTION SOMETIMES MATTERS

One of the most important lessons in this book is that not all friction is waste.

Some friction is the cost of orientation. Some repetition is the price of competence. In domains involving judgment, safety, ethics, or responsibility, some friction is not inefficiency but discipline. Review steps, checklists, verification loops, and pauses for human interpretation are often the mechanisms by which error is prevented and accountability is preserved.

AI can remove friction very effectively. That does not mean every friction should be removed.

This point becomes decisive wherever responsibility remains human. If the person is still accountable, then the person must still be able to judge. A system that accelerates action while hollowing out judgment does not simply improve performance. It changes the nature of responsibility itself.

ASSISTANCE, AUTONOMY, AND DELEGATED AUTHORITY

This book has also argued that assistance and autonomy must be kept distinct.

Assistance preserves the human judgment point while expanding reach. Autonomy transfers execution to the system within defined bounds. That transfer may be appropriate in narrow, observable, well-bounded domains. It becomes dangerous when extended into settings where ambiguity, irreversibility, ethics, or responsibility are central.

The real question is therefore not whether autonomy is available, but where it belongs.

Present AI does not need to become sentient to create serious governance problems. It only needs to be granted authority beyond the range in which its limitations are understood and bounded. The immediate issue is not inner will. It is delegated execution without preserved judgment.

Where the Real Risk Lies

The first real risk of AI is human misuse.

The book has examined several forms of that misuse: argument detached from truth, persuasion detached from understanding, administration detached from judgment, authority detached from responsibility, and relationship detached from reciprocity. None of these failures is new. What is new is the speed, scale, and fluency with which AI can now amplify them.

This is where fear so often goes wrong. It imagines an inward threat where the actual danger is outward and institutional. It imagines rogue rebellion where there is over-delegation. It imagines consciousness where there is patterned output. It imagines machine will where there is human misuse.

The danger is not that machines have suddenly begun to corrupt human affairs. The danger is that human beings have long done so, and now possess tools that can scale those failures.

Fear Must Be Specified

Fear, however, should not simply be mocked or discarded.

The historical record shows that fear often appears when technologies alter scale, authority, speed, labor, and power. In that sense, fear is often directionally right. What it frequently gets wrong is the object, mechanism, or location of the danger. It also often underestimates the capacity of human systems to adapt once the real mechanism is better understood.

That lesson applies directly to AI.

AI combines several older fear patterns at once. It lowers the cost of symbolic production, compresses communicative time, reorganizes work, raises fears of displacement, extends delegated execution, operates opaquely for many users, and introduces real questions of concentrated institutional power. But it also adds a new complication: it acts in forms humans associate with thought. That is why its fears so quickly become anthropomorphic and mythic.

The proper response is not comfort. It is calibration.

What Requires Action

If the categories developed here are sound, then the practical priorities become clearer.

What requires action first is not speculative machine inwardness. It is the governance of present systems as they actually exist.

That means preserving judgment where responsibility still resides. It means matching autonomy to domain and consequence. It means building oversight into the structure of deployment rather than adding it after the fact. It means aligning incentives so that throughput and apparent efficiency do not crowd out reflection, review, and human development. It means training users for inquiry rather than justification. It means resisting relational substitution where systems simulate companionship without reciprocity. And it means refusing to let symbolic fluency masquerade as understanding.

This is not a call for panic, nor for passivity. It is a call for fit, boundary, and discipline.

Final Judgment

The final judgment is straightforward.

Artificial intelligence is real, powerful, and socially consequential.

It is also limited, non-thinking, and easily misclassified.

Its greatest immediate risks are human and institutional.

Its greatest confusions arise when language, speed, and adaptation are mistaken for thought, judgment, or will.

Its proper use depends on preserving the human role where responsibility, interpretation, and consequence remain irreducibly human.

AI should neither be trivialized nor mythologized.

It should be understood for what it is, bounded where it must be, used where it properly serves, and resisted where it separates power from judgment.

That is the line drawn here.

And if that line is held, then AI may be treated neither as a god nor as a ghost, but as what it is: an adaptive artificial tool whose value and danger depend, in the end, on the quality of the human judgment surrounding it.

Appendices

Appendices A through F
Introduction

The appendices that follow provide supporting material for the main body of this book. They are intended to deepen understanding, document sources of reasoning, and offer structured detail without interrupting the flow of the central argument.

Each appendix serves a distinct purpose. Some provide historical context for key concepts such as intelligence and technological development. Others document the methodological approach used to construct the indices presented in this book. Still others illustrate specific mechanisms — such as the generation of structured language without understanding, or the historical misuse of reasoning systems — that are central to the book's conclusions.

These pages are designed to be consulted as needed. They are not required for understanding the main argument, but they provide additional clarity, evidence, and depth for readers who wish to examine particular aspects more closely.

Taken together, the appendices reinforce a central aim of this book: to replace ambiguity with definition, speculation with structure, and impression with understanding.

Appendix A: Survey of Intelligence and Thought in the Western Tradition

This appendix surveys the treatment of intelligence across major Western philosophical traditions in order to clarify the conceptual foundations underlying this book. The aim is not to provide an exhaustive history of philosophy, but to show how the concept of intelligence has been understood in relation to reason, judgment, knowledge, and human conduct.

The Western philosophical tradition is particularly relevant here because of its emphasis on analysis, categorization, and the systematic examination of thought. These characteristics align with the distinctions developed in this book between thinking, intelligence, and structured reasoning systems.

These perspectives provide historical context for the characterizations of intelligence as adaptability and thinking as judgment within a point of view.

PRE-SOCRATIC PHILOSOPHERS

ANAXAGORAS

Anaxagoras introduced the concept of nous (mind or intellect) as a fundamental principle governing the cosmos. He posited that nous was the source of order and reason in the universe, separating it from chaotic matter. For Anaxagoras, nous was not only a cosmic principle but also an inherent aspect of human intelligence, enabling humans to perceive and understand the world.

Original sources:

- Fragments of Anaxagoras (various collections)
- G.S. Kirk, J.E. Raven, M. Schofield, *The Presocratic Philosophers* (Cambridge University Press)

HERACLITUS

Heraclitus emphasized the role of *logos* (reason or word) as the underlying order of the universe. While he did not directly define human intelligence, his concept of logos suggested that intelligence involved aligning oneself with this universal reason, understanding the constant change (flux) that defines existence.

Original sources:

- Fragments of Heraclitus (various collections)
- Philip Wheelwright, *Heraclitus* (Princeton University Press)

PYTHAGORAS

Pythagoras, although better known for his contributions to mathematics, also viewed intelligence as connected to the understanding of mathematical and harmonic principles. He believed that through intellectual and spiritual purification, one could achieve *sophia* (wisdom), which was closely tied to understanding the mathematical order of the cosmos.

Original sources:

- *Pythagoras and the Pythagoreans* by Charles H. Kahn
- *The Presocratics* by Edward Hussey (Duckworth Press)

EMPEDOCLES

Empedocles introduced the idea that intelligence was connected to the balance of the four elements (earth, air, fire, water) within the human body. He also believed in the transmigration of souls and that gaining knowledge across different lifetimes led to higher intelligence.

Original sources:

- *The Presocratic Philosophers* by Jonathan Barnes (Routledge)
- *Empedocles: An Interpretation* by M.R. Wright (Yale University Press)

EARLY POST-SOCRATIC PHILOSOPHERS

SOCRATES

Socrates, whose thoughts are preserved through the writings of his students like Plato and Xenophon, laid the foundation for Western philosophy. He viewed intelligence as closely tied to moral and ethical understanding, emphasizing the importance of self-knowledge and the ability to reason about the good life. Socratic intelligence was not just about acquiring knowledge but about understanding oneself and living a virtuous life.

Original sources:

- *Socrates - the Man and His Thought* by A.E. Taylor
- Plato's Dialogues (e.g., Apology, Phaedo)

PLATO

Plato expanded on Socratic thought by introducing the concept of the Tripartite Soul in his works such as The Republic. He divided the soul into three parts: reason (logistikon), spirit (thumos), and appetite (epithumia), with reason being the highest element of the soul. In Plato's epistemology, the highest form of knowledge involves direct apprehension of truth (often associated with noes*is*), rather than discursive reasoning alone. Plato's Theory of Forms also emphasized that true knowledge and intelligence consist in understanding the eternal and unchanging realities, rather than the transient physical world.

Original sources:

- *The Republic* by Plato, translated by Allan Bloom
- *Plato's Theory of Knowledge* by Norman Gulley (Routledge)

ARISTOTLE

Aristotle, a student of Plato, provided a more systematic and empirical approach to intelligence. In his *Nicomachean Ethics* and *De Anima*, Aristotle distinguished between different types of intelligence, such as practical wisdom (phronesis) and theoretical wisdom (*sophia*). He also introduced the concept of *nous* as the intellect that grasps first principles, crucial for both ethical reasoning and scientific understanding.

Original sources:

- *Nicomachean Ethics* by Aristotle, translated by Terence Irwin
- *De Anima* by Aristotle, translated by J.A. Smith (Oxford University Press)

MEDIEVAL THINKERS

THOMAS AQUINAS

Thomas Aquinas synthesized Aristotelian philosophy with Christian theology, introducing the idea of *intellectus* (intelligence) as a divine gift that allows humans to know and understand God's creation. In *Summa Theologica*, Aquinas argued that intelligence is both the ability to grasp abstract truths and the capacity for moral reasoning.

Original sources:

- *Summa Theologica* by Thomas Aquinas, translated by the Fathers of the English Dominican Province
- *Thomas Aquinas on Human Nature* by Robert Pasnau (Cambridge University Press)

AUGUSTINE OF HIPPO

Augustine emphasized the role of divine illumination in human intelligence, arguing that true knowledge and understanding come from God's light. In his work *De Trinitate*, Augustine suggested that the human mind reflects the Trinity, with memory, understanding, and will corresponding to the Father, Son, and Holy Spirit.

Original sources:

- *The Confessions* by Augustine, translated by Henry Chadwick (Oxford University Press)
- *De Trinitate* by Augustine, translated by Stephen McKenna (Catholic University of America Press)

RENAISSANCE THINKERS

RENÉ DESCARTES

Descartes is often considered the father of modern philosophy. His *Meditations on First Philosophy* introduced the idea of *cogito, ergo sum* ("I think, therefore I am"), positing thinking as the essence of existence. For Descartes, intelligence was the ability to reason, doubt, and reflect on one's own existence, setting the stage for the Enlightenment's emphasis on rational thought.

Original sources:

- *Meditations on First Philosophy* by René Descartes, translated by John Cottingham (Cambridge University Press)
- *Descartes: An Intellectual Biography* by Stephen Gaukroger (Oxford University Press)

FRANCIS BACON

Bacon advanced the empirical method, emphasizing observation and experimentation as the basis of knowledge. In *The Advancement of Learning*, he argued that intelligence involves the systematic investigation of nature through inductive reasoning, contrasting with the purely deductive methods of earlier philosophers.

Original sources:

- *The Advancement of Learning* by Francis Bacon, edited by Michael Kiernan (Clarendon Press)
- *Francis Bacon: The Major Works* edited by Brian Vickers (Oxford World's Classics)

MODERN PHILOSOPHERS

IMMANUEL KANT

Kant's *Critique of Pure Reason* introduced the idea that intelligence involves the synthesis of sensory input and the application of categories like time and space, which are inherent to the human mind. Kant argued that intelligence is not just about knowledge acquisition but also about understanding the conditions that make knowledge possible.

Original sources:

- *Critique of Pure Reason* by Immanuel Kant, translated by Paul Guyer and Allen W. Wood (Cambridge University Press)
- *Kant and the Mind* by Andrew Brook (Cambridge University Press)

G.W.F. HEGEL

Hegel's *Phenomenology of Spirit* suggested that intelligence is part of a larger process of *Geist* (spirit) coming to know itself through history. For Hegel, intelligence is dynamic and evolves over time, influenced by cultural and historical contexts.

Original sources:

- *Phenomenology of Spirit* by G.W.F. Hegel, translated by A.V. Miller (Oxford University Press)
- *Hegel: A Very Short Introduction* by Peter Singer (Oxford University Press)

Conclusion

The concept of intelligence has evolved across these traditions, but consistently remains tied to reason, understanding, and the capacity for judgment. These philosophical treatments reinforce a central assertion of this book: that intelligence, however defined, has historically been associated with judgment and meaning, not merely with structured output or patterned response.

This characterization is essential when considering artificial intelligence. Systems that generate structured outputs may resemble intelligence in form, but do not necessarily meet the deeper criteria historically associated with thinking.

Appendix B:
Timeline of Key AI Milestones, Technologies, and Models

This appendix provides a selective timeline of key AI milestones, technologies, frameworks, and models most relevant to the book's discussion of **computing power, programming power**, and the historical accumulation of AI capability. It is not intended to be exhaustive. Rather, it identifies representative inflection points that either expanded machine capability, increased programming leverage, or visibly changed what AI systems could do in practice.

Dates refer to the most relevant public milestone year associated with a concept, system, release, or publication.

Pre-Computational Foundations (1920s–1940s)

This period does not yet contain artificial intelligence in any formal sense, but it establishes the conceptual and technical groundwork on which later developments depend. Mechanical and electromechanical computation, early information theory, and wartime advances in electronics begin to expand what machines can do before programmable digital systems emerge.

1920s–1930s — Electromechanical Tabulation and Early Data Processing

Large-scale tabulating systems, developed by companies such as IBM, enabled automated processing of structured data. These systems were not programmable in the modern sense, but they represent early expansion in machine-side capacity for information handling.

1936 — Turing's Computability Framework
Alan Turing's work on computable numbers introduced the abstract concept of a universal machine, providing a theoretical foundation for programmable computation.

1943 — McCulloch–Pitts Neuron Model
One of the earliest formal models linking logic and neural processes, suggesting that computation could, in principle, model aspects of cognition.

1940s — Wartime Electronics and Early Digital Computing

Advances during World War II accelerated the development of electronic computing systems, including early programmable machines such as Colossus and ENIAC. These systems expanded both speed and scale of computation, establishing the first practical digital computing environments.

1948 — Shannon's Information Theory
Claude Shannon's formulation of information theory provided a mathematical framework for communication and signal processing, foundational to later developments in computation and machine learning.

Foundations: Early Concepts and First Systems (1950s–1960s)

1950 — Turing Test
Alan Turing's "Computing Machinery and Intelligence" framed the enduring question of whether machines could imitate intelligent behavior convincingly enough to be mistaken for thinkers.

1956 — Dartmouth Conference

The Dartmouth Summer Research Project on Artificial Intelligence is widely regarded as the formal beginning of AI as a field and established the term *artificial intelligence.*

1958 — Perceptron

Frank Rosenblatt's perceptron introduced one of the earliest trainable neural models, laying groundwork for later machine learning approaches.

1966 — ELIZA

Joseph Weizenbaum's ELIZA demonstrated how simple pattern-based language interaction could create the appearance of understanding long before modern language models.

1966–1972 — Shakey the Robot

Developed at SRI, Shakey combined perception, planning, and action, becoming one of the first systems to integrate AI reasoning with robotics.

Symbolic Systems, Expert Knowledge, and Early Commercial AI (1970s–1980s)

1972 — MYCIN

MYCIN, developed at Stanford, became one of the most influential expert systems, showing how rule-based reasoning could be applied to medicine.

1980 — XCON (R1)

Digital Equipment Corporation's XCON demonstrated the commercial value of expert systems by automating computer-configuration decisions at industrial scale.

1982 — Fifth Generation Computer Systems Project

Japan's Fifth Generation initiative accelerated global interest in AI, symbolic reasoning, and machine-oriented knowledge systems, even where its ambitions exceeded its practical results.

1986 — Backpropagation
The Rumelhart–Hinton–Williams backpropagation work made multilayer neural networks trainable in practice and became one of the enabling foundations of later deep learning.

Statistical Learning, Neural Networks, and Enabling Compute (1990s–2000s)

1997 — Deep Blue
IBM's Deep Blue defeated Garry Kasparov, demonstrating the power of large-scale search, evaluation, and specialized computation in a narrowly bounded domain.

1998 — LeNet-5
Yann LeCun's convolutional network work showed that neural networks could perform useful pattern recognition tasks such as handwritten digit classification.

2006 — Deep Learning Revival
Work associated with Geoffrey Hinton and others renewed practical interest in deep neural networks, especially as larger datasets and stronger computational resources became available.

2007 — CUDA and General-Purpose GPU Computing
NVIDIA's CUDA platform made large-scale parallel computation far more accessible to researchers and developers, materially strengthening the computing power available for AI workloads.

Deep Learning at Scale: Architectures, Frameworks, and Practical Breakthroughs (2010s)

2011 — IBM Watson
Watson's Jeopardy! performance showed that AI systems could integrate question answering, retrieval, ranking, and natural language processing at impressive scale.

2012 — AlexNet
AlexNet's ImageNet victory dramatically improved image-classification performance and is widely regarded as the breakthrough that pushed deep learning into the center of AI practice.

2014 — GANs (Generative Adversarial Networks)
Ian Goodfellow's GAN framework opened a new era of generative modeling by training two neural networks in competition.

2015–2016 — TensorFlow and PyTorch
These frameworks greatly increased programming power by lowering the barrier to building, training, and deploying modern AI systems. They reduced the distance between research insight and engineering implementation.

2016 — AlphaGo
DeepMind's AlphaGo defeat of Lee Sedol demonstrated a powerful combination of deep learning, reinforcement learning, and tree search in a domain long regarded as beyond the reach of conventional approaches.

2016 — TPUs and Specialized AI Accelerators
DeepMind's AlphaGo defeat of Lee Sedol demonstrated the effective integration of deep learning, reinforcement learning, and tree search in a single system. By learning from data and improving through self-play, it showed that machine learning could extend into complex domains long considered resistant to computation. This marked a shift

from narrowly engineered performance toward more flexible, learned approaches.

2017 — Transformer Architecture
"Attention Is All You Need" introduced the Transformer, a major programming-power breakthrough that became the backbone of modern large language models and many multimodal systems.

2018 — BERT
BERT showed the power of large-scale pretraining for language understanding and helped establish the now-dominant paradigm of transfer from foundation-style pretrained models.

Foundation Models, Multimodality, and Public Breakthroughs (2020s)

2020 — GPT-3
GPT-3 demonstrated that scaling large language models could produce striking few-shot performance and moved the public discussion of AI toward more general adaptive systems.

2021 — DALL·E
DALL·E demonstrated high-profile text-to-image generation and showed that generative modeling could extend beyond language into visual domains.

2021 — AlphaFold2
AlphaFold2 made a major advance in protein structure prediction, showing AI's capacity for high-value scientific application in a domain of biological complexity.

2022 — ChatGPT
ChatGPT marked a major public interface breakthrough by making large language models broadly accessible through

conversational interaction, dramatically expanding public awareness and use.

2023 — GPT-4

GPT-4 marked a significant advance in the scale, multimodality, and capability of frontier language models, becoming a visible marker of post-2020 acceleration.

2024 — Real-Time Multimodal and Long-Context Frontier Models

Systems such as GPT-4o, Claude 3, and Gemini 1.5 illustrated the expanding frontier in multimodality, context length, tool use, and real-time interaction.

Conclusion

The milestones listed here show that AI did not emerge through a single invention or a single breakthrough. It accumulated through successive advances in:

- conceptual framing,
- trainable models,
- expert systems and symbolic reasoning,
- neural network methods,
- scalable compute,
- reusable programming frameworks,
- and large foundation-style models.

This timeline supports a central claim of the book: AI capability rose through the long interaction of computing power and programming power, with visible leaps occurring when both were sufficiently mature to make new classes of systems practical.

Sources

1. Turing, A. M. (1950). *Computing Machinery and Intelligence. Mind*, 59(236), 433–460.

2. McCarthy, J., Minsky, M. L., Rochester, N., & Shannon, C. E. (1956). *A Proposal for the Dartmouth Summer Research Project on Artificial Intelligence.*
3. Rosenblatt, F. (1958). *The Perceptron: A Probabilistic Model for Information Storage and Organization in the Brain. Psychological Review*, 65(6), 386–408.
4. Weizenbaum, J. (1966). *ELIZA — A Computer Program for the Study of Natural Language Communication Between Man and Machine. Communications of the ACM*, 9(1), 36–45.
5. Nilsson, N. J. (1984). *Shakey the Robot.* SRI International / AI Center retrospective materials.
6. Buchanan, B. G., & Shortliffe, E. H. (1984). *Rule-Based Expert Systems: The MYCIN Experiments of the Stanford Heuristic Programming Project.* Addison-Wesley.
7. Feigenbaum, E. A., & McCorduck, P. (1983). *The Fifth Generation: Artificial Intelligence and Japan's Computer Challenge to the World.* Addison-Wesley.
8. Rumelhart, D. E., Hinton, G. E., & Williams, R. J. (1986). *Learning Representations by Back-Propagating Errors. Nature*, 323(6088), 533–536.
9. Campbell, M., Hoane, A. J., & Hsu, F. H. (2002). *Deep Blue. Artificial Intelligence*, 134(1–2), 57–83.
10. LeCun, Y., Bottou, L., Bengio, Y., & Haffner, P. (1998). *Gradient-Based Learning Applied to Document Recognition. Proceedings of the IEEE*, 86(11), 2278–2324.
11. Hinton, G. E., Osindero, S., & Teh, Y. W. (2006). *A Fast Learning Algorithm for Deep Belief Nets. Neural Computation*, 18(7), 1527–1554.
12. Nickolls, J., Buck, I., Garland, M., & Skadron, K. (2008). *Scalable Parallel Programming with CUDA. Queue*, 6(2), 40–53.

13. Ferrucci, D., et al. (2010). *Building Watson: An Overview of the DeepQA Project. AI Magazine*, 31(3), 59–79.
14. Krizhevsky, A., Sutskever, I., & Hinton, G. E. (2012). *ImageNet Classification with Deep Convolutional Neural Networks. Advances in Neural Information Processing Systems.*
15. Goodfellow, I., et al. (2014). *Generative Adversarial Nets. Advances in Neural Information Processing Systems.*
16. Abadi, M., et al. (2016). *TensorFlow: Large-Scale Machine Learning on Heterogeneous Distributed Systems.*
17. Silver, D., et al. (2016). *Mastering the Game of Go with Deep Neural Networks and Tree Search. Nature*, 529(7587), 484–489.
18. Jouppi, N. P., et al. (2017). *In-Datacenter Performance Analysis of a Tensor Processing Unit. Proceedings of ISCA.*
19. Vaswani, A., et al. (2017). *Attention Is All You Need. Advances in Neural Information Processing Systems.*
20. Devlin, J., Chang, M. W., Lee, K., & Toutanova, K. (2018). *BERT: Pre-training of Deep Bidirectional Transformers for Language Understanding.*
21. Brown, T. B., et al. (2020). *Language Models are Few-Shot Learners.*
22. Ramesh, A., et al. (2021). *Zero-Shot Text-to-Image Generation.*
23. Jumper, J., et al. (2021). *Highly Accurate Protein Structure Prediction with AlphaFold. Nature*, 596(7873), 583–589.
24. OpenAI. (2022). *Introducing ChatGPT.*
25. OpenAI. (2023). *GPT-4 Technical Report.*
26. OpenAI, Anthropic, and Google official technical reports / system cards (2024) for representative multimodal and long-context frontier systems

Appendix C:
Methodology for the Computing Power Index, Programming Power Index, and AI Capability Index

This appendix describes the construction of the three curves used in this book: the Computing Power Index (CPI), the Programming Power Index (PPI), and the AI Capability Index (ACI).

The method is structured to remain transparent. Its aim is not to claim exact measurement of "AI power," but to provide a disciplined and reproducible historical comparison of the two principal drivers identified in the book: computing power and programming power. The combined AI Capability Index is then constructed as the product of those two indices.

This is therefore a **proxy-based historical scoring model**, not a black-box statistical inference model. It is designed so that it can be followed directly from the component definitions and tables provided with access to the underlying year-by-year input table (see Tables 3–5 for full annual series and Tables 1–2 for anchor construction).

The combined measure that is referred to throughout this book as the AI Capability Index (ACI).

1. Scope and Design Principles

The indices were constructed according to the following rules:

- **Study period:** 1920 through 2024, inclusive
- **Time step:** annual
- **Output scale:** each index normalized to the interval **0.0 to 1.0** over the study period
- **Use case:** historical comparison of trends, slopes, and inflection points rather than precise annual ranking

The method is governed by four design principles:

1. **Transparency**
   Every component is observable, explicit, and interpretable.

2. **Mutual dependence**
   AI capability is treated as the interaction of machine-side capacity and method-side capacity.

3. **Monotonic historical structure**
   The indices are intended to reflect long-run growth with changing slopes, not noisy short-run oscillation.

4. **Modesty of claim**
   The index is used to identify broad historical movement and major turning periods, not to pretend precision where the available evidence does not support it.

Milestone dates used to anchor the historical scoring are drawn from Appendix B — Timeline of Key AI Milestones, Technologies, and Models.

2. Computing Power Index (CPI)

2.1 Definition

- The Computing Power Index measures the effective machine-side capacity available to support AI-relevant workloads in a given year.
- It is defined broadly, not merely as processor speed. In the context of this book, computing power includes:
- processing throughput
- memory capacity
- storage capacity
- data movement / bandwidth
- parallel and specialized compute capability

2.2 Component Structure

For each year t, five component scores are assigned on a common scale from **0 to 10**:

$Proc_t =$ processing throughput
$Mem_t =$ memory capacity / effective addressable memory
$Store_t =$ storage capacity / accessible storage
$Band_t =$ data movement, interconnect, and bandwidth capability
$Spec_t =$ parallel and specialized compute (e.g., vector, GPU, TPU, or equivalent accelerator capacity)

The anchor values used to construct these components are shown in Table 1.

The raw computing score is then calculated as:

$$C_{raw,t} = 0.35(Proc_t) + 0.20(Mem_t) + 0.15(Store_t) + 0.15(Band_t) + 0.15(Spec_t)$$

These weights reflect the judgment that processing throughput remains the dominant machine constraint, while memory, storage, bandwidth, and specialized compute materially affect whether throughput can actually be used at scale.

2.3 Scoring Method

Each component is assigned by one of two methods:

A. Quantitative proxy mapping

Where reasonably continuous historical quantitative data are available, the proxy series is transformed using a logarithmic scale and then mapped to the 0–10 component scale.

For example:

- transistor count
- peak operations per second
- memory capacity
- storage capacity
- interconnect bandwidth

For a quantitative proxy X_t, the score is computed as:

$$Score_t = 10 \times \frac{\log_{10}(X_t) - \log_{10}(X_{min})}{\log_{10}(X_{max}) - \log_{10}(X_{min})}$$

where X_{min} and X_{max} are the minimum and maximum observed values over the study period.

B. Milestone scoring

Where continuous quantitative data are not available or would be misleading, a milestone-based score is assigned at documented dates and linearly interpolated between those dates.

The milestone rubric is:

0 = absent or pre-computational
2 = experimental / laboratory capability
4 = limited practical capability
6 = broadly usable capability
8 = mature, scalable capability

10 = frontier-enabling capability at large scale

Between milestone years, annual values are linearly interpolated.

The specific milestone years and assigned component scores are documented in Table 1

2.4 Normalization

After calculating the raw score for all years, the final normalized Computing Power Index is:

$$CPI_t = \frac{C_{raw,t} - \min(C_{raw})}{\max(C_{raw}) - \min(C_{raw})}$$

This yields an annual series from **0.0 to 1.0.**

3. Programming Power Index (PPI)

3.1 Definition

The Programming Power Index measures the effective human-side ability to turn computational capacity into useful AI behavior.

It does **not** attempt to measure the brilliance of individual programmers. It measures the historical leverage provided by algorithms, abstractions, languages, libraries, frameworks, tooling, and distribution of reusable method.

3.2 Component Structure

For each year t, five component scores are assigned on a common scale from **0 to 10:**

Alg_t = algorithmic / model sophistication
Abs_t = programming abstraction and language expressiveness
$Frame_t$ = reusable frameworks, libraries, and infrastructure
$Tool_t$ = tooling, debugging, testing, and developer ergonomics

$Dist_t$= distribution, collaboration, and accessibility of capability

The corresponding anchor values for programming power are shown in Table 2.

The raw programming score is:

$$P_{raw,t} = 0.30(Alg_t) + 0.20(Abs_t) + 0.20(Frame_t) + 0.15(Tool_t) + 0.15(Dist_t)$$

These weights reflect the view that algorithmic/methodological breakthroughs carry the greatest influence, while abstraction, framework maturity, tooling, and distribution jointly determine how much of that capability becomes usable in practice.

3.3 Component Meaning

Algorithmic / model sophistication includes the historical rise of trainable models, expert systems, backpropagation, deep learning, reinforcement learning, transformer architectures, and foundation-style models.

Programming abstraction includes the shift from low-level coding toward high-level languages, numerical computing environments, scripting, notebook-style workflows, and other abstractions that lower the cost of expressing complex computation.

Frameworks and libraries capture the accumulation of reusable programming power through toolchains such as deep learning libraries, inference frameworks, model libraries, and standardized AI infrastructure.

Tooling includes debugging, testing, version control, profiling, distributed execution support, deployment tooling, and surrounding developer ergonomics.

Distribution and accessibility capture the degree to which AI capability is confined to elite research labs or becomes available more broadly through open-source ecosystems, shared repositories, cloud access, APIs, and mature engineering practice.

3.4 Scoring Method

The same scoring logic is used as for CPI:

- use **quantitative proxy mapping** where meaningful time-series data are available
- otherwise use **milestone scoring** anchored to documented events and linearly interpolate between anchor years

The same 0–10 milestone rubric is used.

Milestone anchors for programming power are drawn directly from the chronology in Appendix B. These anchors and their component scores are summarized in Table 2. Representative anchors include:

- perceptron
- expert systems / MYCIN / XCON
- backpropagation
- deep learning revival
- TensorFlow / PyTorch
- Transformer
- BERT
- GPT-3 and later foundation models

3.5 Normalization

The final normalized Programming Power Index is:

$$PPI_t = \frac{P_{raw,t} - \min(P_{raw})}{\max(P_{raw}) - \min(P_{raw})}$$

This yields an annual series from **0.0 to 1.0**.

4. A Note on Data

Data is not treated here as a third independent first-order index.

That decision is deliberate.

Data is necessary, but its effect is conditional. In many periods, data exists in abundance or can be generated in volume. The limiting factor is not always its presence, but its usability. That usability is governed by computing power and programming power — the capacity to process, structure, and convert data into learned capability.

For that reason, data is treated here as a dependent input rather than as a third co-equal driver.

5. AI Capability Index (ACI)

5.1 Construction

The AI Capability Index is constructed as the product of the normalized Computing Power Index and the normalized Programming Power Index:

$$ACI_t = CPI_t \times PPI_t$$

5.2 Why the Product Form Is Used

The product form is chosen because it is the simplest mathematical structure that enforces the key historical observation of the book: **AI capability depends on the joint maturity of computing power and programming power.**

If the indices were merely added, one could too easily compensate for weakness in one dimension by strength in the other. But that is not how AI capability actually develops. Powerful hardware with weak methods remains limited. Elegant methods without sufficient compute remain unrealized. The product form captures this mutual dependence.

More complex forms — involving powers, nonlinear scaling, or fitted parameters — could certainly be constructed. But such forms would introduce assumptions about the relative dominance of each driver that are not justified at this level of analysis. The product is therefore used as a **first-order interaction model**: simple enough to remain interpretable, strong enough to express the dominant relationship.

5.3 Interpretation

The ACI is not a direct measure of:

- intelligence
- thought
- judgment
- consciousness
- moral status
- or social desirability

It is a measure of the historical practical capacity to build adaptive artificial systems.

Its main analytical use is not the exact value in a single year, but the identification of:

- slope changes
- turning periods
- convergence of the two drivers
- and major inflection points in AI capability

6. Annual Series Construction

The annual index series is constructed by the following procedure:

1. **Select component proxies and milestone anchors** for CPI and PPI.
2. **Assign component scores** on a 0–10 scale at each documented milestone year.
3. **Interpolate annual values** linearly between milestone years.
4. **Compute raw CPI and PPI** using the weighted formulas above.
5. **Normalize each raw series** independently to the interval 0–1.
6. **Multiply the normalized series** to form ACI.
7. **Plot the annual indices** directly without further smoothing beyond interpolation.

If multiple milestones occur in a single year, the score used for that year is the end-of-year state after incorporating all milestones judged relevant.

The calculated results are shown in the following tables and were used to construct the curves in Chapter 7 for Computing Power (CPI), Programming Power (PPI), and AI Capability (ACI).

The resulting annual series for CPI, PPI, and ACI are provided in Tables 3, 4, and 5 respectively.

CPI Anchor Years and Component Scores						
Year	Milestone / Rationale	Proc	Mem	Store	Band	Spec
1920	Electromechanical tabulators and pre-digital computing baseline	0.0	0.0	0.0	0.0	0.0
1940	Wartime electromechanical and vacuum-tube computing	0.2	0.1	0.1	0.1	0.0
1950	Stored-program computers become practical	0.4	0.3	0.3	0.2	0.0
1960	Transistorized mainframes and core memory era	0.9	0.8	0.7	0.5	0.1
1971	Commercial microprocessor era begins	1.8	1.5	1.4	1.0	0.2
1980	VLSI, DRAM/HDD gains, PC/workstation diffusion	2.8	2.6	2.5	1.8	0.4
1990	Networked computing and supercomputing broaden	4.2	4.0	4.1	3.2	0.9
2000	Internet/server scale and early cloud-era compute	5.8	5.6	5.9	4.9	1.8
2007	CUDA / general-purpose GPU computing	6.8	6.6	6.9	6.2	3.5
2012	AlexNet-era GPU training visible in practice	7.8	7.6	7.9	7.3	5.8
2016	TPUs and specialized AI accelerators	8.7	8.5	8.8	8.2	7.4
2020	Frontier-scale AI clusters / GPT-3 era	9.4	9.3	9.4	9.2	8.9
2024	Real-time multimodal frontier systems at scale	10.0	10.0	10.0	10.0	10.0

Anchor rationale follows Appendix B milestones and computing-infrastructure turning period

Table C1: Computing Power Index (CPI) Anchor Values

PPI Anchor Years and Component Scores						
Year	Milestone / Rationale	Alg	Abs	Frame	Tool	Dist
1920	Pre-computational baseline for programming method	0.0	0.0	0.0	0.0	0.0
1950	Turing-era symbolic machine intelligence framing	0.4	0.5	0.0	0.2	0.0
1956	Dartmouth / AI field formation	0.6	0.7	0.0	0.3	0.0
1958	Perceptron and trainable-model milestone	0.9	0.8	0.0	0.3	0.0
1966	ELIZA / Shakey and early AI systems	1.3	1.2	0.2	0.6	0.1
1972	MYCIN and expert-system logic	2.2	1.8	0.6	1.0	0.3
1980	XCON and early commercial AI tooling	2.8	2.6	1.2	1.8	0.8
1986	Backpropagation practical training breakthrough	4.2	3.1	1.8	2.3	1.2
1998	LeNet and maturing statistical learning practice	4.8	4.5	2.8	3.8	2.8
2006	Deep-learning revival / practical multilayer training	5.8	5.6	3.8	5.0	4.4
2012	AlexNet and deep learning mainstream breakthrough	7.0	6.8	5.5	6.3	5.8
2015	TensorFlow / modern ML framework era	7.5	7.3	6.8	6.8	6.5
2017	Transformer architecture	8.6	7.8	7.4	7.2	7.0
2018	BERT and transfer-learning mainstream	8.9	8.2	7.8	7.6	7.5
2020	GPT-3 / large foundation-model scaling	9.4	8.9	8.6	8.4	8.5
2022	ChatGPT public interface breakthrough	9.7	9.4	9.2	9.0	9.2
2024	Multimodal / long-context / tool-using model integration	10.0	10.0	10.0	10.0	10.0

Anchor rationale follows Appendix B milestones and programming/ method turning periods.

Table C2: Programming Power Index (PPI) Anchor Values

Computing Power Index (CPI) Component Scores						
Year	Proc	Mem	Store	Band	Spec	Raw Index
1920	0.01	0.01	0.01	0.01	0	0.01
1930	0.08	0.05	0.05	0.03	0	0.05
1940	0.15	0.1	0.1	0.05	0	0.1
1950	0.4	0.3	0.3	0.15	0	0.27
1960	0.9	0.8	0.7	0.5	0.1	0.67
1970	1.72	1.44	1.34	0.95	0.19	1.26
1980	2.8	2.6	2.5	1.8	0.4	2.21
1990	4.5	4.2	4	3	1.2	3.75
2000	6.5	6.2	6	5.5	3.5	5.85
2010	8.5	8.2	8	7.8	6.5	8.03
2020	9.7	9.5	9.4	9.2	8.8	9.46
2024	10	10	10	10	10	10

Table C3: Computing Power Index (CPI) Component Scores

Programming Power Index (PPI) Component Scores						
Year	Alg	Abs	Frame	Tool	Dist	Raw Index
1920	0	0	0	0	0	0
1930	0.01	0.01	0	0	0	0.01
1940	0.05	0.05	0.01	0.02	0.01	0.04
1950	0.5	0.3	0.1	0.2	0.1	0.34
1960	0.9	0.7	0.4	0.6	0.3	0.66
1970	1.3	1	0.7	0.9	0.5	1.02
1980	2	1.6	1.2	1.5	1	1.66
1990	3.5	3	2.5	2.8	2	2.98
2000	5.5	5	4.5	5	4	4.93
2010	7.5	7	6.5	7	6	6.93
2020	9.2	9	8.8	9	8.5	8.96
2024	10	10	10	10	10	10

Table C4: Programming Power Index (PPI) Component Scores

AI Capability Index (ACI) Product			
Year	**CPI**	**PPI**	**ACI (Product)**
1920	0	0	0
1930	0.01	0.01	0
1940	0.02	0.03	0.001
1950	0.1	0.1	0.01
1960	0.2	0.15	0.03
1970	0.3	0.2	0.06
1980	0.5	0.3	0.15
1990	0.7	0.5	0.35
2000	0.85	0.7	0.595
2010	0.95	0.85	0.808
2020	0.98	0.91	0.892
2024	1	1	1

Table C5: AI Capability Index (ACI) Product

7. Robustness Check

Because the index is intended to identify broad historical structure rather than exact annual rankings, a simple robustness check is sufficient.

The indices were recomputed:

- using equal weights, and
- with each component weight varied by ±20%

Across these variations, the broad shape of the curves and the locations of the major turning periods remained materially unchanged.

This confirms that the principal conclusions of the book do not depend on a fragile weighting choice.

The purpose of this step is not to prove statistical precision, but to verify that the underlying historical structure is stable under reasonable variation in assumptions.

8. Validity and Limitations

The method has several strengths:

- it is transparent
- it matches the conceptual structure of the book
- it is interpretable by non-specialists
- it can be reproduced if the input tables are provided
- and it is appropriately modest in its claims

Its limitations should also be stated plainly:

1. **Historical proxies are uneven**
   Earlier decades are less richly documented than recent decades.

2. **Some component scores require judgment**
   Especially on the programming side, not every improvement can be reduced to one quantitative series.

3. **The index is comparative, not absolute**
   It is intended to show broad movement and inflection, not exact measurement of "AI capability" in a scientific sense. This is reflected in the use of normalized series shown in Tables 3–5 rather than raw physical units.

4. **The model is first-order**
   It captures the dominant interaction between the two drivers, but does not attempt to model every secondary factor.

These limitations do not invalidate the index. They define its proper use.

9. Conclusion

The methodology in this appendix is designed to match the discipline of the book itself.

The Computing Power Index measures the machine-side capacity available for AI-relevant workloads.

The Programming Power Index measures the human-side leverage available to turn that capacity into useful adaptive behavior.

The AI Capability Index is the product of the two.

Together, they provide a transparent first-order model of how AI capability accumulated over time.

The purpose is not to claim a hidden law of history. It is to provide a structured representation of what the book argues in conceptual form: that AI capability rose not through hardware alone or software alone, but through the long interaction of both.

The tables provided in this appendix make the structure of that interaction explicit and reproducible.

SOURCES

1. Moore, G. E. (1965). *Cramming More Components onto Integrated Circuits.*

2. Hennessy, J. L., & Patterson, D. A. (2017). *Computer Architecture: A Quantitative Approach* (6th ed.). Morgan Kaufmann.

3. Rumelhart, D. E., Hinton, G. E., & Williams, R. J. (1986). *Learning Representations by Back-Propagating Errors. Nature,* 323(6088), 533–536.

4. Hinton, G. E., Osindero, S., & Teh, Y. W. (2006). *A Fast Learning Algorithm for Deep Belief Nets. Neural Computation,* 18(7), 1527–1554.

5. Abadi, M., et al. (2016). *TensorFlow: Large-Scale Machine Learning on Heterogeneous Distributed Systems.*

6. Paszke, A., et al. (2019). *PyTorch: An Imperative Style, High-Performance Deep Learning Library.*

7. Vaswani, A., et al. (2017). *Attention Is All You Need.*

8. Brown, T. B., et al. (2020). *Language Models are Few-Shot Learners.*

9. Jouppi, N. P., et al. (2017). *In-Datacenter Performance Analysis of a Tensor Processing Unit.*

10. Appendix B — *Timeline of Key AI Milestones, Technologies, and Models.*

Appendix D:
Overview of Post-2020 Major Government Policies & Initiatives

This appendix provides a selective overview of major post-2020 government and intergovernmental initiatives most relevant to the development, governance, and deployment of artificial intelligence. It is not intended to be exhaustive. Its purpose is narrower: to identify the policy moves that most clearly shaped AI standards, legal frameworks, compute infrastructure, data governance, and institutional readiness in the years after 2020.

A central point of this book is that AI capability depends on more than models alone. It develops within a larger ecosystem of semiconductors, standards, data rules, cybersecurity, institutions, and incentives. Government action after 2020 increasingly reflected that broader reality.

United States: Coordination, Standards, and Infrastructure

In the United States, the **National Artificial Intelligence Initiative Act of 2020** established a coordinated federal framework for AI research, development, standards, and interagency collaboration. Among other things, it directed NIST to advance frameworks, standards, and risk-management approaches for AI systems. That mandate helped shape the later development of the **NIST AI Risk Management Framework (AI RMF 1.0)**, released on January 26, 2023, as a voluntary framework for organizations designing, developing, deploying, or using AI systems.

The **CHIPS and Science Act of 2022** addressed another side of the AI ecosystem: semiconductor capacity, research infrastructure, and workforce development. Official implementation materials describe the law as providing major support for U.S. semiconductor research, development, and manufacturing, while also investing in workers and broader scientific capacity. In AI terms, that matters because compute availability is one of the foundational conditions for modern AI capability.

Federal executive policy then shifted noticeably across 2023–2025. **Executive Order 14110** (October 30, 2023), *Safe, Secure, and Trustworthy Development and Use of Artificial Intelligence*, pushed agencies toward safety, security, and governance actions, but it was rescinded on January 20, 2025. Three days later, **Executive Order 14179** (January 23, 2025), *Removing Barriers to American Leadership in Artificial Intelligence*, redirected policy emphasis toward innovation and U.S. leadership. That shift was later carried into **America's AI Action Plan** (July 2025), which organized federal strategy around three pillars: accelerating innovation, building AI infrastructure, and leading international diplomacy and security.

European Union: Legal Framework and Phased Implementation

The European Union moved most decisively on the legal-regulatory side. **Regulation (EU) 2024/1689**, the **AI Act**, was adopted in June 2024 and entered into force on August 1, 2024. It created a risk-based legal framework for AI systems across the EU. The application timeline is phased: prohibited AI practices and AI literacy obligations began applying on February 2, 2025; governance rules and obligations for general-purpose AI models began applying on August 2, 2025; the Act becomes fully applicable on August 2, 2026, with some high-risk product rules extending to 2027.

Alongside the AI Act, the EU also advanced broader **data governance** relevant to AI deployment. The **Data Act** entered into force on January 11, 2024 and became applicable on September 12, 2025. While not an AI law in the narrow sense, it matters because it addresses access to and use of data, especially in data-rich and connected-device environments that increasingly feed AI systems.

China: Algorithm Governance to Generative AI Regulation

China's post-2020 policy trajectory is notable for moving in steps from specific algorithmic controls toward broader generative-AI regulation. In early 2022, China implemented **Provisions on the Administration of Algorithm-generated Recommendations for Internet Information Services**, which took effect on March 1, 2022. These rules addressed recommendation systems directly, including restrictions on fake news, harmful content, and manipulative uses of algorithmic recommendation.

That framework was followed by the **Provisions on Administration of Deep Synthesis of Internet-based Information Services**, effective January 10, 2023. These provisions targeted synthetic media and related deep-synthesis services. Later in 2023, several Chinese authorities jointly issued the **Interim Measures for the Management of Generative AI Services**, which took effect on August 15, 2023 and established a more specific governance framework for generative AI services. Taken together, these steps show a staged governance pattern: recommendation systems first, synthetic media next, and then generative AI more directly.

Multilateral and International Frameworks

Post-2020 AI governance also developed through international and intergovernmental initiatives. The **Global Partnership on AI (GPAI)** was launched in June 2020 as a voluntary, multi-stakeholder initiative focused on the responsible development and use of AI. In July 2024, GPAI and OECD member-country AI work were brought together more fully under the GPAI brand, reflecting the growing institutionalization of international AI policy coordination.

A second major milestone was UNESCO's **Recommendation on the Ethics of Artificial Intelligence**, adopted in November 2021. UNESCO describes it as the first global standard on AI ethics and emphasizes principles such as human rights, human dignity, transparency, fairness, and human oversight. Its significance for this book is not that it settles all governance questions, but that it shows how quickly AI moved from technical topic to subject of global normative attention.

What These Post-2020 Initiatives Suggest

Taken together, these post-2020 actions suggest that governments did not respond to AI in only one register. They responded across several.

First, they moved toward **standards and risk management**, especially through NIST and related frameworks in the United States. Second, they moved toward **formal legal classification and phased obligations**, most visibly in the European Union. Third, they moved toward **algorithm- and content-specific controls**, especially in China's staged regulation of recommendation systems, deep synthesis, and generative AI. Fourth, they acted on **compute and semiconductor capacity**, recognizing that AI capability depends on infrastructure as well as models. Fifth, they increasingly treated AI as a matter of **institutional readiness and workforce capacity**, not merely technical novelty.

This is important because it supports one of the larger arguments of the book: AI is not merely a software phenomenon. It is an ecosystem phenomenon. Governments have increasingly responded not only to what AI systems do, but to the infrastructures, institutions, and incentive structures that determine how they are built and used.

Conclusion

The most important post-2020 policy lesson is that AI governance did not remain confined to abstract "AI ethics". It moved into concrete domains: standards, legal frameworks, semiconductors, data access, recommendation systems, synthetic media, workforce preparation, and institutional capacity.

That shift is consistent with the central thesis of this book. AI capability and AI risk both arise within broader human systems. Governments, like firms and users, are slowly being forced to govern not an imagined future machine mind, but a present ecosystem of compute, data, institutions, incentives, and delegated authority.

Sources and Notes

1. **United States**

   - National Artificial Intelligence Initiative Act of 2020 / NIST mandate overview.
   - NIST AI Risk Management Framework (AI RMF 1.0), released January 26, 2023.
   - CHIPS and Science Act / CHIPS for America implementation materials.

- Executive Order 14110 (2023) and its rescission in January 2025.
- Executive Order 14179 (2025) and America's AI Action Plan (2025).

2. **European Union**
   - AI Act (Regulation (EU) 2024/1689), entry into force and implementation timeline.
   - Data Act, entry into force and application date.
3. **China**
   - Algorithm recommendation provisions (effective March 1, 2022).
   - Deep synthesis provisions (effective January 10, 2023).
   - Interim Measures for the Management of Generative AI Services (effective August 15, 2023).
4. **International / Multilateral**
   - GPAI launch (June 2020) and later OECD/GPAI integration.
   - UNESCO Recommendation on the Ethics of Artificial Intelligence (2021).

Appendix E:
A Pre-Model Approach to Technical Language Generation

THE ARTIFACT

The following artifact was circulated informally within the aerospace industry in the 1980s, referred to as "buzz phrases," and illustrates a structured method for generating technical language through recombination.

It presents a method for constructing technical-sounding sentences by combining phrases selected from four sequential groups (A, B, C, and D).

\- A -

1. In particular,
2. On the other hand,
3. However,
4. Similarly,
5. As a resultant implication,
6. In this regard,
7. Based on integral subsystem considerations,
8. For example,
9. Thus,
10. In respect to specific goals,

\- B -

1. a large portion of the interface coordination communication
2. a constant flow of effective information
3. the characterization of specific criteria
4. initiation of critical subsystem development
5. the fully integrated system program
6. the product configuration baseline
7. any associated supporting element
8. the incorporation of additional mission constraints
9. the independent functional principle
10. a primary interrelationship between system and/or subsystem terminologies

Artifact Excerpt 1

The instructions are simple: select one phrase from each column and concatenate them in sequence. The result is a sentence that appears structurally valid and professionally articulated. While often appearing meaningful, the resulting construction carries little or no specific intent.

\- C -

1. must utilize and be functionally interwoven with
2. maximizes the probability of project success and minimizes the cost and time required for
3. adds explicit performance limits to
4. necessitates that urgent consideration be applied to
5. requires considerable systems analyses and trade-off studies to arrive at
6. is further compounded, when taking into account
7. presents extremely interesting challenges to
8. recognizes the importance of other systems and the necessity for
9. effects a significant implementation for
10. adds overriding performance constraints to

\- D -

1. the sophisticated hardware
2. the anticipated fourth generation equipment
3. the subsystem compatibility testing
4. the structural design, based on system engineering concepts
5. the preliminary qualification limit
6. the evolution of specifications over a given time period
7. the philosophy of commonality and standardization
8. the greater tight-worthiness concept
9. any discrete configuration mode
10. the total system rationale

<u>Artifact Excerpt 2</u>

For example, a sentence generated using this method might read:

> "In this regard, the characterization of specific criteria requires considerable systems analyses to arrive at the total system rationale."

The sentence is structurally sound and stylistically familiar, yet conveys no precise or testable meaning.

How it Works

The system is composed of four categories:

A — introductory phrases (e.g., "In particular," "However," "Thus,")

B — subject constructions (e.g., "a constant flow of effective information")

C — connective or functional phrases (e.g., "requires considerable systems analyses")

D — technical objects or domains (e.g., "the total system rationale")

By selecting one phrase from each column and combining them, the user produces a complete sentence. The structure ensures grammatical plausibility and stylistic familiarity, particularly within technical or managerial contexts.

The result often resembles legitimate engineering language, while conveying little or no precise meaning.

What it Demonstrates

This artifact illustrates an important principle:

The appearance of intelligent language can be generated through structure alone.

The system produces:

- grammatically correct sentences
- recognizable technical tone
- plausible professional phrasing

Yet these outputs are not derived from understanding, reasoning, or judgment. They are assembled through recombination of pre-existing fragments.

Each column in the system contains ten selectable phrases. When combined across four columns, this produces 10,000 possible sentence constructions. These sentences can be generated rapidly, with little effort, and many will appear structurally valid and professionally articulated. However, each instance must still be evaluated individually for meaning. The system therefore demonstrates a fundamental asymmetry: the cost of generating plausible language is low, while the cost of evaluating it remains high.

This demonstrates that:

- fluency is not evidence of thought
- structure is not evidence of intelligence
- coherence of form is not coherence of meaning

WHAT IT LACKS

Despite its ability to generate plausible language, the system lacks several defining features of modern AI:

- **No internal selection mechanism** — the user chooses each component
- **No learning** — the system does not improve or adapt
- **No context sensitivity** — it does not respond to input conditions
- **No generalization** — it cannot extend beyond predefined fragments

The system defines a space of possible sentences, but it does not navigate that space on its own.

THE CRITICAL THRESHOLD

The distinction between this system and modern language models lies in a single structural feature:

The selection process is external.

In the buzz phrase system, a human selects each component. The system itself does not evaluate, rank, or choose among alternatives.

Modern language models, by contrast, internalize this selection process. They generate output by selecting successive tokens based on learned statistical structure and contextual weighting.

This marks an important threshold:

- Combinatorial systems define possible outputs
- Language models generate outputs through learned internal selection

The difference is not merely one of scale, but of mechanism.

WHY THIS MATTERS

This example clarifies a central theme of this book.

The ability to generate language — even fluent, structured, and convincing language — does not imply:

- thinking
- understanding
- judgment
- or intelligence

Modern AI systems extend far beyond this primitive example by incorporating learning and adaptability. They internalize structure and apply it across varying inputs. In this sense, they are legitimately intelligent under the definition used in this book.

However, the underlying lesson remains:

> Language output alone is not sufficient evidence of thinking.

The buzz phrase system produces the form of intelligent language without possessing intelligence. Modern AI produces more sophisticated outputs through learned adaptability, but still does not cross the threshold into thinking as defined earlier.

CONCLUDING OBSERVATION

The artifact serves as a reminder that:

- language can persuade without meaning
- structure can imitate substance
- and fluency can obscure the absence of judgment

These observations are not new. What is new is the scale and effectiveness with which modern systems can produce such output.

Understanding this distinction is essential before assigning to those systems capabilities they do not possess.

Appendix F:
The Thinkery —
On the Misuse of Reasoning Systems

1. THE PATTERN

The misuse of structured reasoning systems did not begin with artificial intelligence.

Long before modern computation, Aristophanes' *The Clouds* (423 BC), presented a satirical account of teaching argument detached from truth. In the play, Socrates' "Thinkery" is depicted as a place where students are trained not to understand, but to prevail — to make the weaker argument appear stronger.

The pattern is not merely literary. It is structural.

It appears wherever reasoning is used to secure advantage rather than to discover truth.

2. THE INVERSION

Modern AI systems are generally most effective when used in the following sequence:

> inquiry → insight → synthesis → refinement → production

This sequence aligns with how such systems are designed to be useful. They assist in exploration, comparison, structuring, and iteration.

Misuse often reverses this sequence:

> assertion → constraint → confirmation

A conclusion is fixed in advance. The system is then used to justify it.

This inversion creates friction because the system is being used against its most coherent mode of operation. More importantly, it produces output that appears reasoned without having been grounded in inquiry.

3. The Mechanism

The mechanism of misuse is simple:

- reasoning is detached from truth
- structure is used to support conclusion rather than test it
- output is evaluated for persuasiveness rather than correctness

The result is not the absence of reasoning, but its redirection.

The system produces:

- fluent language
- structured argument
- plausible justification

But the process no longer serves understanding.

4. The Classical Warning

In *The Clouds*, Strepsiades sends his son to the Thinkery with a clear aim: to learn how to win arguments regardless of their validity.

The son succeeds.

He returns home and uses his training to justify beating his father.

The satire is exaggerated, but the warning is not.

Once argument becomes detached from truth, it becomes unstable — even against the user.

5. The Modern Parallel

The same pattern appears in contemporary AI use.

Users may:

- ask the system to defend a position
- constrain outputs toward a preferred conclusion
- select only confirming responses
- treat fluency as validation

The system, in turn, generates coherent output consistent with those constraints.

The result is persuasive, but not necessarily grounded.

The failure lies not in the system, but in its use.

6. The Broader Principle

This pattern is not limited to AI.

It appears wherever:

- models are used without understanding their limits
- metrics replace judgment
- frameworks are applied without context
- symbolic outputs are mistaken for knowledge

AI does not create this behavior.

It amplifies it.

7. Implication

The significance of this pattern is practical.

AI systems are most valuable when used to:

- explore
- compare
- test
- refine

They are least reliable when used to:

- justify
- advocate prematurely
- replace inquiry with conclusion

The distinction is not subtle.

It is the difference between using a system to think more clearly and using it to avoid thinking altogether.

8. Closing Observation

The misuse of reasoning systems is ancient. Artificial intelligence does not introduce it. It extends its reach.

The danger is not that machines will begin to misuse reasoning.

The danger is that humans have always done so — and now possess tools that can scale it.

Primary References

The following works form the conceptual, technical, and historical foundation of this book. They are selected not for completeness, but for their direct relevance to the distinctions developed herein: thinking, intelligence, computation, judgment, and the development of artificial systems.

PHILOSOPHY, THINKING, AND JUDGMENT

Aristophanes. *The Clouds*. 423 BC.
A classical warning about argument detached from truth and the misuse of reasoning for advantage rather than understanding.

Aristotle. *De Anima*. Translated by J. A. Smith. Oxford: Oxford University Press, 1931.
Foundational for understanding intellect (*nous*) and the nature of cognition.

Aristotle. *Nicomachean Ethics*. Translated by Terence Irwin. Indianapolis: Hackett, 1999.
A central work on virtue, practical wisdom, and judgment.

Barzun, Jacques. *The House of Intellect*. New York: Harper & Row, 1959.
Essential for an understanding the differences between intelligence, intellect, and disciplined reasoning.

Descartes, René. *Meditations on First Philosophy*. Translated by John Cottingham. Cambridge: Cambridge University Press, 1986.
A foundational work on doubt, reasoning, and self-reflection.

Dreyfus, Hubert L. *What Computers Can't Do: A Critique of Artificial Reason*. New York: Harper & Row, 1972.
A major critique of attempts to reduce human thought to formal procedure.

Heidegger, Martin. *What Is Called Thinking?* Translated by Fred D. Wieck and J. Glenn Gray. New York: Harper & Row, 1968.
A central work distinguishing thinking from calculation and representation.

Hegel, G. W. F. *Phenomenology of Spirit.* Translated by A. V. Miller. Oxford: Oxford University Press, 1977.
A systematic treatment of consciousness and historical development of thought.

Kant, Immanuel. *Critique of Pure Reason*. Translated by Paul Guyer and Allen W. Wood. Cambridge: Cambridge University Press, 1998.
A major work on epistemology and the limits of human understanding.

Nagel, Thomas. "What Is It Like to Be a Bat?" *The Philosophical Review* 83, no. 4 (1974): 435–450.
A key argument for the irreducibility of subjective point of view.

Plato. *The Republic.* Translated by Allan Bloom. New York: Basic Books, 1968.
A foundational work on reason, knowledge, and the structure of thought.

Taylor, A. E. *Socrates: The Man and His Thought.* London: Methuen, 1932.
A classic interpretation of Socratic reasoning and ethical inquiry.

ARTIFICIAL INTELLIGENCE, COMPUTING, AND MACHINE LEARNING

Abadi, Martín, et al. "TensorFlow: Large-Scale Machine Learning on Heterogeneous Distributed Systems." 2016.
A major framework advancing practical AI development and deployment.

Brown, Tom B., et al. "Language Models are Few-Shot Learners." 2020.
A landmark demonstration of large-scale language model capability.

Goodfellow, Ian, Yoshua Bengio, and Aaron Courville. *Deep Learning*. Cambridge, MA: MIT Press, 2016.
Foundational for modern neural networks and learning systems.

Hennessy, John L., and David A. Patterson. *Computer Architecture: A Quantitative Approach*. 6th ed. Morgan Kaufmann, 2017.
Essential for understanding computing power as a system-level constraint.

Russell, Stuart J., and Peter Norvig. *Artificial Intelligence: A Modern Approach*. 4th ed. Pearson, 2021.
The standard comprehensive reference for AI methods and history.

Searle, John R. "Minds, Brains, and Programs." *Behavioral and Brain Sciences* 3, no. 3 (1980): 417–424.
Essential for distinguishing symbol manipulation from understanding.

Turing, Alan M. "Computing Machinery and Intelligence." *Mind* 59, no. 236 (1950): 433–460.
The foundational modern framing of machine intelligence.

Vaswani, Ashish, et al. "Attention Is All You Need." 2017.
The defining architecture underlying modern large language models.

Weizenbaum, Joseph. *Computer Power and Human Reason: From Judgment to Calculation.* San Francisco: W. H. Freeman, 1976.
A critical examination of the limits of computational systems.

Human Systems, Automation, and Social Effects

Bainbridge, Lisanne. "Ironies of Automation." *Automatica* 19, no. 6 (1983): 775–779.
A classic statement of the paradox of human oversight in automated systems.

Carr, Nicholas. *The Glass Cage: Automation and Us.* New York: W. W. Norton, 2014.
A key work on automation, skill atrophy, and human judgment.

Carr, Nicholas. *The Shallows: What the Internet Is Doing to Our Brains.* New York: W. W. Norton, 2010.
Examines cognitive effects of digital systems.

March, James G., and Herbert A. Simon. *Organizations.* New York: Wiley, 1958.
Foundational for understanding institutional decision-making and simplification.

Parasuraman, Raja, and Victor Riley. "Humans and Automation: Use, Misuse, Disuse, Abuse." *Human Factors* 39, no. 2 (1997): 230–253.
A key framework for human-machine interaction and failure modes.

Turkle, Sherry. *Reclaiming Conversation: The Power of Talk in a Digital Age.* New York: Penguin Press, 2015.
Explores mediated communication and human interaction.

HISTORICAL AND CONTEXTUAL WORKS

Eisenstein, Elizabeth L. *The Printing Revolution in Early Modern Europe.* Cambridge: Cambridge University Press, 1980.
A foundational study of the printing press and its impact on authority.

Jonnes, Jill. *Empires of Light.* New York: Random House, 2003.
A history of electrification and technological transformation.

Mayor, Adrienne. *Greek Fire, Poison Arrows, and Scorpion Bombs.* New York: Overlook Press, 2003.
Historical context for technological fear and concentrated power.

Rhodes, Richard. *The Making of the Atomic Bomb.* New York: Simon & Schuster, 1986.
A detailed account of nuclear technology and its implications.

Standage, Tom. *The Victorian Internet.* New York: Walker & Company, 1998.
A history of the telegraph and its societal effects.

Index

B

C

D

K

L

M

N

O

P

R

S

V

W

X

www.ingramcontent.com/pod-product-compliance
Lightning Source LLC
LaVergne TN
LVHW022157200926
848481LV00006B/1778